T0254964

Lecture Notes in Computer Science 13722

More information about this series at https://link.springer.com/bookseries/558

Suheyla Cetin-Karayumak · Daan Christiaens · Matteo Figini · Pamela Guevara · Tomasz Pieciak · Elizabeth Powell · Francois Rheault (Eds.)

Computational Diffusion MRI

13th International Workshop, CDMRI 2022 Held in Conjunction with MICCAI 2022 Singapore, Singapore, September 22, 2022 Proceedings

Editors
Suheyla Cetin-Karayumak
Harvard Medical School
Boston, MA, USA

Daan Christiaens
KU Leuven
Leuven, Belgium

Matteo Figini
University College London
London, UK

Pamela Guevara
Universidad de Concepción
Concepción, Chile

Tomasz Pieciak
Universidad de Valladolid
Valladolid, Spain

Elizabeth Powell
University College London
London, UK

Francois Rheault
Université de Sherbrooke
Sherbrooke, QC, Canada

ISSN 0302-9743 ISSN 1611-3349 (electronic)
Lecture Notes in Computer Science
ISBN 978-3-031-21205-5 ISBN 978-3-031-21206-2 (eBook)
https://doi.org/10.1007/978-3-031-21206-2

This Springer imprint is published by the registered company Springer Nature Switzerland AG
The registered company address is: Gewerbestrasse 11, 6330 Cham, Switzerland

Preface

It is our privilege to introduce the proceedings of this year's International Workshop on Computational Diffusion MRI (CDMRI 2022). Since its conception over ten years ago, CDMRI has been organized each year as a satellite event of the Medical Image Computing and Computer Assisted Intervention (MICCAI) conference, bringing together researchers from across the globe to discuss the latest advances in the acquisition, analysis, and application of diffusion MRI (dMRI).

Diffusion MRI has transformed our ability to non-invasively probe tissue microstructure and function over the last four decades, with applications ranging from the mapping of neural pathways to assisting with tumor classification to aiding neurosurgical planning and beyond. However, to continue progression in this field and access even more complex microstructural features, richer and more sophisticated acquisition protocols are required that may not be feasible in clinical settings. In recent years, machine learning techniques have begun to offer the possibility of achieving equivalent information from smaller, more clinically feasible datasets, and as such are becoming increasingly prevalent in the dMRI literature. The submissions to CDMRI 2022 are reflective of this trend: over 80% of the proffered papers proposed machine learning strategies to improve either dMRI image processing or data storage. These numbers also highlight the synergy with the machine-learning-oriented MICCAI community. Readers of these proceedings can expect to find contributions on topics such as susceptibility distortion correction using semi-supervised deep learning, microstructural model fitting using self-supervised learning, and data compression using neural networks.

However, post-hoc data augmentation techniques are not infallible. This year, the CDMRI workshop hosted the MICCAI challenge entitled 'Quality Augmentation in Diffusion MRI for Clinical Studies: Validation in Migraine' (QuaD22), which asked the pertinent question 'Are we losing relevant quantitative clinical information when generating high-quality images with artificial intelligence techniques?' The QuaD22 organizers tasked participants with generating DTI-related parameters from a dMRI migraine dataset acquired with 21 diffusion gradient directions that were of a similar high-quality to parameters obtained from a dataset with 61 gradient directions, with the aim of evaluating whether it was possible to detect the same significant differences between episodic migraine and chronic migraine patients in the reduced acquisition scenario as in the fully-sampled situation.

This workshop would not have been possible without the dedication of the Program Committee (listed below), who, through a double-blind peer review process, ensured the highest standard was achieved. Of the 13 submissions received, four were accepted without revision, eight were accepted after revision, and one was rejected; each submission was reviewed by at least two members of the Program Committee. We extend our gratitude to everyone involved in the Program Committee. Finally, we wish to thank our keynote speakers (listed below), who delivered a series of truly enlightening lectures.

September 2022

Suheyla Cetin-Karayumak
Daan Christiaens
Matteo Figini
Pamela Guevara
Tomasz Pieciak
Elizabeth Powell
Francois Rheault

Organization

Program Committee Chairs

Suheyla Cetin-Karayumak	Harvard Medical School, USA
Daan Christiaens	KU Leuven, Belgium
Matteo Figini	University College London, UK
Pamela Guevara	Universidad de Concepción, Chile
Tomasz Pieciak	Universidad de Valladolid, Spain
Elizabeth Powell	University College London, UK
Francois Rheault	Université de Sherbrooke, Canada

Program Committee

Nagesh Adluru	University of Wisconsin–Madison, USA
Maryam Afzali	Cardiff University, UK
Suyash P. Awate	Indian Institute of Technology (IIT) Bombay, India
Dogu Baran Aydogan	Aalto University, Finland
Yaël Balbastre	Massachusetts General Hospital, USA
Pietro Bontempi	University of Verona, Italy
Ryan P. Cabeen	University of Southern California, USA
Maxime Chamberland	Donders Institute for Brain, Cognition and Behaviour, The Netherlands
Alessandro Daducci	University of Verona, Italy
Gabriel Girard	EPFL, Switzerland
Ting Gong	University College London, UK
Muge Karaman	University of Illinois at Chicago, USA
Emma Muñoz-Moreno	IDIBAPS, Spain
Marco Palombo	Cardiff University, UK
Marco Pizzolato	Technical University of Denmark, Denmark
Alonso Ramirez-Manzanares	CIMAT A.C., Mexico
Simona Schiavi	University of Genoa, Italy
Thomas Schultz	University of Bonn, Germany
Jelle Veraart	NYU School of Medicine, USA
Ye Wu	University of North Carolina at Chapel Hill, USA
Pew-Thian Yap	University of North Carolina at Chapel Hill, USA
Fan Zhang	Harvard Medical School, USA

Keynote Speakers

Suyash P. Awate	Indian Institute of Technology (IIT), Bombay, India
Amy Howard	University of Oxford, UK
Susie Y. Huang	Martinos Center for Biomedical Imaging, USA
Evren Özarslan	Linköping University, Sweden
Andrew Zalesky	University of Melbourne, Australia

Contents

Data Preprocessing

Slice Estimation in Diffusion MRI of Neonatal and Fetal Brains in Image and Spherical Harmonics Domains Using Autoencoders 3
Hamza Kebiri, Gabriel Girard, Yasser Alemán-Gómez, Thomas Yu, András Jakab, Erick Jorge Canales-Rodríguez, and Meritxell Bach Cuadra

Super-Resolution of Manifold-Valued Diffusion MRI Refined by Multi-modal Imaging 14
Tyler A. Spears and P. Thomas Fletcher

Lossy Compression of Multidimensional Medical Images Using Sinusoidal Activation Networks: An Evaluation Study 26
Matteo Mancini, Derek K. Jones, and Marco Palombo

Correction of Susceptibility Distortion in EPI: A Semi-supervised Approach with Deep Learning 38
Antoine Legouhy, Mark Graham, Michele Guerreri, Whitney Stee, Thomas Villemonteix, Philippe Peigneux, and Hui Zhang

The Impact of Susceptibility Distortion Correction Protocols on Adolescent Diffusion MRI Measures 50
Talia M. Nir, Julio E. Villalón-Reina, Paul M. Thompson, and Neda Jahanshad

Signal Representations

Diffusion MRI Fibre Orientation Distribution Inpainting 65
Zihao Tang, Xinyi Wang, Mariano Cabezas, Arkiev D'Souza, Fernando Calamante, Dongnan Liu, Michael Barnett, Chenyu Wang, and Weidong Cai

Fitting a Directional Microstructure Model to Diffusion-Relaxation MRI Data with Self-supervised Machine Learning 77
Jason P. Lim, Stefano B. Blumberg, Neil Narayan, Sean C. Epstein, Daniel C. Alexander, Marco Palombo, and Paddy J. Slator

Stepwise Stochastic Dictionary Adaptation Improves Microstructure Reconstruction with Orientation Distribution Function Fingerprinting 89
Patryk Filipiak, Timothy Shepherd, Lee Basler, Anthony Zuccolotto, Dimitris G. Placantonakis, Walter Schneider, Fernando E. Boada, and Steven H. Baete

How Can Spherical CNNs Benefit ML-Based Diffusion MRI Parameter Estimation? 101
Tobias Goodwin-Allcock, Jason McEwen, Robert Gray, Parashkev Nachev, and Hui Zhang

Tractography and WM Pathways

DC^{2}U-Net: Tract Segmentation in Brain White Matter Using Dense Criss-Cross U-Net 115
Haoran Yin, Pengbo Xu, Hui Cui, Geng Chen, and Jiquan Ma

Clustering in Tractography Using Autoencoders (CINTA) 125
Jon Haitz Legarreta, Laurent Petit, Pierre-Marc Jodoin, and Maxime Descoteaux

Tractometric Coherence of Fiber Bundles in DTI 137
Rick Sengers, Tom Dela Haije, Andrea Fuster, and Luc Florack

Author Index 149

Data Preprocessing

Slice Estimation in Diffusion MRI of Neonatal and Fetal Brains in Image and Spherical Harmonics Domains Using Autoencoders

Hamza Kebiri[1,2](✉), Gabriel Girard[1,2,3], Yasser Alemán-Gómez[1], Thomas Yu[3,4], András Jakab[5,6], Erick Jorge Canales-Rodríguez[3], and Meritxell Bach Cuadra[1,2]

[1] Department of Radiology, Lausanne University Hospital and University of Lausanne, Lausanne, Switzerland

[2] CIBM Center for Biomedical Imaging, Lausanne, Switzerland

hamza.kebiri@unil.ch

[3] Signal Processing Laboratory 5 (LTS5), École Polytechnique Fédérale de Lausanne (EPFL), Lausanne, Switzerland

[4] Advanced Clinical Imaging Technology, Siemens Healthineers International AG, Lausanne, Switzerland

[5] Center for MR Research, University Children's Hospital Zurich, Zurich, Switzerland

[6] Neuroscience Center Zurich, University of Zurich, Zurich, Switzerland

Abstract. Diffusion MRI (dMRI) of the developing brain can provide valuable insights into the white matter development. However, slice thickness in fetal dMRI is typically high (i.e., 3–5 mm) to freeze the in-plane motion, which reduces the sensitivity of the dMRI signal to the underlying anatomy. In this study, we aim at overcoming this problem by using autoencoders to learn unsupervised efficient representations of brain slices in a latent space, using raw dMRI signals and their spherical harmonics (SH) representation. We first learn and quantitatively validate the autoencoders on the developing Human Connectome Project pre-term newborn data, and further test the method on fetal data. Our results show that the autoencoder in the signal domain better synthesized the raw signal. Interestingly, the fractional anisotropy and, to a lesser extent, the mean diffusivity, are best recovered in missing slices by using the autoencoder trained with SH coefficients. A comparison was performed with the same maps reconstructed using an autoencoder trained with raw signals, as well as conventional interpolation methods of raw signals and SH coefficients. From these results, we conclude that the recovery of missing/corrupted slices should be performed in the signal domain if the raw signal is aimed to be recovered, and in the SH domain if diffusion tensor properties (i.e., fractional anisotropy) are targeted. Notably, the trained autoencoders were able to generalize to fetal dMRI data acquired using a much smaller number of diffusion gradients and a lower b-value, where we qualitatively show the consistency of the estimated diffusion tensor maps.

S. Cetin-Karayumak et al. (Eds.): CDMRI 2022, LNCS 13722, pp. 3–13, 2022.
https://doi.org/10.1007/978-3-031-21206-2_1

Keywords: Super-resolution · Autoencoders · Spherical harmonics · Diffusion tensor imaging · Pre-term · Fetal · Brain · MRI

1 Introduction

Neonatal and fetal brain development involves complex cerebral growth and maturation both for gray and white matter [4,10]. Diffusion MRI (dMRI) has been widely employed to study this developmental process *in vivo*, including neonates and fetuses [16,18,28]. As the diffusion weighted signal is sensitive to the displacement of water molecules, several models have been proposed for estimating the underlying anatomy such as diffusion tensor imaging (DTI) or spherical deconvolution methods [2,6,32]. The accuracy of these models is dependant on the angular and spatial resolution of the acquisitions that is typically limited for the neonate and fetal subjects [19,22]. Stochastic motion and low signal-to-noise ratio (SNR) due to the small size of the developing brain often translate to degraded images with low spatial resolution. Additionally, slice thickness in fetal dMRI is typically high, varying between 3–5 mm, to freeze the in-plane motion, and hence reduces the sensitivity of the dMRI signal to the underlying anatomy. This highlights the need for methods to interpolate or synthesize new slices that were either (1) corrupted because of motion or (2) acquired using anisotropic voxel sizes. Interpolation is often performed either at scanner level or in post-processing [19], and has been demonstrated to be relevant for raw signal recovery and for subsequent analysis such as tractography [11]. Similarly, super-resolution (SR) methods that aim at increasing dMRI resolution can be applied at the acquisition-reconstruction level [27,29] or at post-processing [5,7,12]. The latter used *supervised learning* methods, which require high resolution training data that is often unavailable for the developing brain. Additionally, these methods focus on enhancing the resolution homogeneously over all dimensions and were not assessed for anisotropic voxels, commonly acquired for fetuses and neonates [19,22]. Additionally to the raw dMRI signal interpolation, other representations such as Spherical Harmonics (SH) could be of interest. SH are a combination of smooth orthogonal basis functions defined on the surface of a sphere able to represent spherical signals, such as the dMRI signal acquired using uniformly distributed gradient directions [13,15]. Previous work used deep learning methods to map the SH coefficients from one shell to another [20,24]. However, no prior work, to the best of our knowledge, relies on the SH decomposition to enhance the spatial image resolution.

In this study, we have used *unsupervised learning* to extend the application of autoencoders for through-plane super-resolution [21,30] in the image domain to spherical harmonics domain where we synthesize SH coefficients of *missing* slices. As such, our network has access to both angular and spatial information. In contrast to training with non-DWI volumes [21], we have additionally trained a second network on spherical averaged dMRI images to complement and compare its performance in relation to the SH trained network. Moreover, we have compared both methods to conventional interpolation methods both using raw dMRI signals and their SH representation. The comparison was performed both on the raw dMRI signal; and on fractional anisotropy (FA) and mean diffusivity

(MD) maps derived from the estimated diffusion tensors. Finally, we verified that the SH networks trained on pre-term data successfully generalized to fetal images, where we present the coherence of the synthesized slices.

2 Methodology

2.1 Materials

Neonatal Data - The developing Human Connectome Project (dHCP) data[1] were acquired in a 3T Philips Achieva scanner in a multi-shell scheme ($b \in \{0, 400, 1000, 2600\}$ s/mm^2). Details on acquisition parameters can be found in [17]. The data was denoised, motion and distortion corrected [3] and has a final resolution of $1.17 \times 1.17 \times 1.5$ mm^3 in a FOV of $128 \times 128 \times 64$ mm^3. In addition to $b = 0$ s/mm^2 images (b0), we have selected the corresponding 88 volumes with $b = 1000$ s/mm^2 (b1000) from all pre-term subjects (31) defined with less than 37 gestational weeks (GW) ([29.3, 37.0], mean = 35.5). In the anatomical dataset, brain tissue labels and masks [26] were provided.

Fetal Data - The fetal data were acquired with the approval of the ethics committee. Acquisitions were performed at 1.5T (GE Healthcare) with a single shot echo planar imaging sequence (TE = 63 ms, TR = 2200 ms) using $b = 700$ s/mm^2 (b700) and 15 directions. The acquisition FOV was $256 \times 256 \times 14 - 22$ mm^3 for a resolution of $1 \times 1 \times 4 - 5$ mm^3. Three axial and one coronal acquisitions were performed for each subject. Four subjects were used in our study: two of 35 and 29 GW where three axial volumes were used, and two young subjects of 24 GW where one axial volume was used. We have only used axial acquisitions to avoid any confounding factor due to interpolation in the registration that would be needed between the orthogonal orientations. Volumes were corrected for noise [34], bias-field inhomogeneities [33] and distortions [1,25] and did not require any motion correction.

2.2 Model

Network Architecture - Our network is composed of four blocks in the encoder and four blocks in the decoder, where each block consists of two layers of 3×3 convolutions, a batch normalization and an Exponential Linear Unit (ELU) activation function [9]. After each block of the encoder, a 2×2 average pooling operation was performed and the number of feature maps was doubled after each layer. Hence starting from 32 feature maps to 256 while three additional 3×3 convolutions were added in the last block with 512, 256 and M feature maps respectively, $M \in \{16, 32, 64, 128\}$. The last M feature maps were considered as the latent space of our autoencoder. The decoder goes back to original input dimensions by means of either 3×3 transposed convolutions with strides of 2 or by 2×2 nearest neighbor interpolations (mutually exclusive), where the number of feature maps decreases by two after each layer from 512 to 32. A last

[1] http://www.developingconnectome.org/data-release/second-data-release/.

1×1 convolution with sigmoid activation function was performed to generate the predicted image.

Training - Using the same architecture, we have trained three networks, with different inputs: b0 images (*b0-net*), average b1000 (*Avg-b1000-net*) (see *Raw signal networks* subsection) and a maximum SH order (L_{max}) of 4 (*SH4-net*) (see *Spherical harmonics networks* subsection). Input images were first normalized to the range $[0, 1]$ by $x = \frac{x - x_{min}}{x_{max} - x_{min}}$ where x_{min} and x_{max} are the minimum and maximum intensities respectively in a given slice. All networks were trained using an Nvidia GeForce RTX 3090 GPU in the TensorFlow framework (version 2.4.1) with Adam optimizer [23] for 200 epochs using mean squared error loss function, a batch size of 32 and a learning rate of 5×10^{-5}. The validation was performed on 15% of the training data. The number of feature maps of the latent space was optimized using Keras-tuner [8] and the checkpoint with the minimal validation loss was finally selected for inference.

Raw Signal Networks - While *b0-net* was trained using b0 images, *Avg-b1000-net* was trained on average b1000 images, as training directly on individual b1000 images did not consistently converge [21]. We have thus trained *Avg-b1000-net* on average b1000 images with the aim of increasing the SNR and reducing variability. The average was computed over n randomly selected volumes, $n \in \{3, 6, 15, 30, 40\}$. Empirically, higher n means a lower risk of network divergence, at the cost of increased smoothness/risk of losing image detail. Therefore n must be tuned. In the end, *b0-net* was used to infer b0 images whereas *Avg-b1000-net* was used to infer b1000 volumes.

Spherical Harmonics Network - We have fit SH representations by using $L_{max} = 4$ to the dMRI signal using Dipy [14] and fed the resulting 15 SH coefficients, slice by slice, to *SH4-net*. Let us note that we preliminary computed the mean squared error difference with respect to the ground truth data when estimating SH and projecting back to original grid from SH bases of $L_{max} \in \{4, 6, 8\}$. As differences were relatively low between them (9.80, 8.64 and 9.95 for $L_{max} \in \{4, 6, 8\}$ respectively, scale $\times 10^{-4}$) and we aim at further testing on fetal data (where only 15 DWI are available) we selected to stick in what follows to $L_{max} = 4$.

Inference in Neonates - For all networks (*b0-net*, *Avg-b1000-net* and *SH4-net*), nested cross validation was performed where the 31 subjects were split into 8 folds. For each subject and each volume in the testing set, we removed N intermediate slices, $N \in \{1, 2\}$ that were considered as the ground truth we aim to predict. Using the two adjacent slices, we input each separately to the encoder part of the network to get the M latent feature maps. These feature maps were averaged using an equal weighting for $N = 1$ and a $\{\frac{1}{3}, \frac{2}{3}\}$, $\{\frac{2}{3}, \frac{1}{3}\}$ weighting for $N = 2$ (Fig. 1). The missing slices were then recovered by using the decoder part from the resulting latent feature maps. The output of the network was then mapped back to the range of input intensities. This was performed using histogram matching (using cumulative probability distributions) between the network output as a source image and the (weighted) average of the two adjacent input slices as a reference image. Finally, the histogram matched output of *SH4-net* was projected back to the original grid of 88 directions to recover the dMRI signal in the image domain.

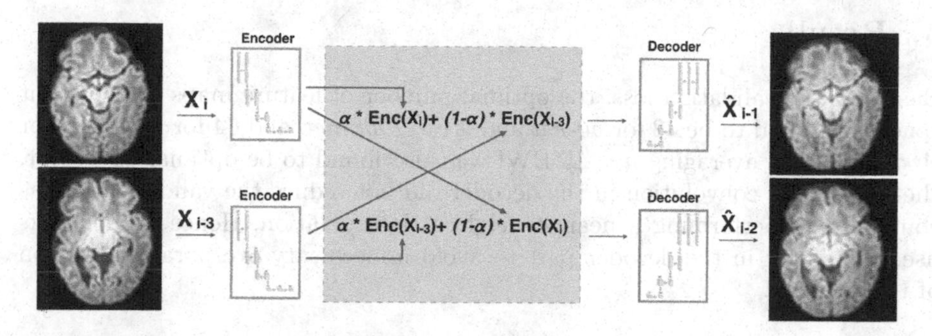

Fig. 1. Inference for two adjacent slices of the first coefficient of SH-L_{max} order 4 illustrated for the case of $N = 2$ where $\alpha = \frac{2}{3}$.

Evaluation in Neonates - The inferred slices of *Avg-b1000-net* were compared to conventional interpolations, namely trilinear, tricubic and B-spline of 5th order [1,31]. The comparison was performed separately for one and two missing slices ($N \in \{1, 2\}$) using the mean squared error (MSE). As all interpolation baselines produce similar results with a slight overperformance for the linear method (for $N = 2$, MSE of 0.003164, 0.003204 and 003211 for linear, cubic and B-spline respectively), the former was chosen for further comparison with autoencoders. The two networks were additionally compared for FA and MD maps that were extracted from the diffusion tensors , as estimated in Dipy [14]. The DTI fit used the synthesized b0 by *b0-net.* The linear baseline was further compared with *SH4-net* and with the signal recovered from the same interpolation of the SH coefficients. The comparison was also extended for DTI maps (FA, MD). To compute them, DTI fit of *SH4-net* relied on the b0 as synthesized by *b0-net,* and the linear SH4 used corresponding linear interpolated b0. All comparisons were done using MSE for FA and MD maps in white matter, cortical gray matter, and corpus callosum. Moreover, we have fit SH representations of the ground truth signal by using $L_{max} = 4$ which were compared after projecting back to the original grid of 88 gradient unit vectors to the original DWI signal, separately for ($N \in \{1, 2\}$). This was considered as the lower bound error of *SH4-net.*

Application to Fetal DWI - After fitting the SH coefficients with L_{max}=4 to the fetal data. We have used *SH4-net,* i.e., trained on pre-term neonates to infer SH coefficients of middle ($N \in \{1, 2\}$) slices of fetal subjects. The inference was performed in a similar manner as for neonates (Fig. 1). Cropping of fetal images to 128 × 128 voxels was necessary before feeding them to the encoder. Then, we generated the diffusion tensor based on this new DWI signal and b0 using *b0-net,* and visually assessed the consistency of the new slices in MD and

FA maps for the four subjects. Only qualitative evaluation was performed for fetal enhancement because of the lack of ground truth.

3 Results

Based on the validation loss, the optimal number of feature maps in the latent space was found to be 32 for *b0-net* and *Avg-b1000-net*, and 64 for *SH4-net*. For *Avg-b1000-net*, averaging $n = 15$ DWI was also found to be optimal. Moreover, the transposed convolution in the decoder did not reduce the validation loss as compared to performing a nearest neighbor interpolation. Hence all networks used the latter in the decoder part to avoid unnecessary overparameterization of the network.

3.1 DWI Assessment

Autoencoder average b1000 trained network (*Avg-b1000-net*) produces superior results compared to linear interpolation (Fig. 2). The difference is higher for the case of two slices removed ($N = 2$).

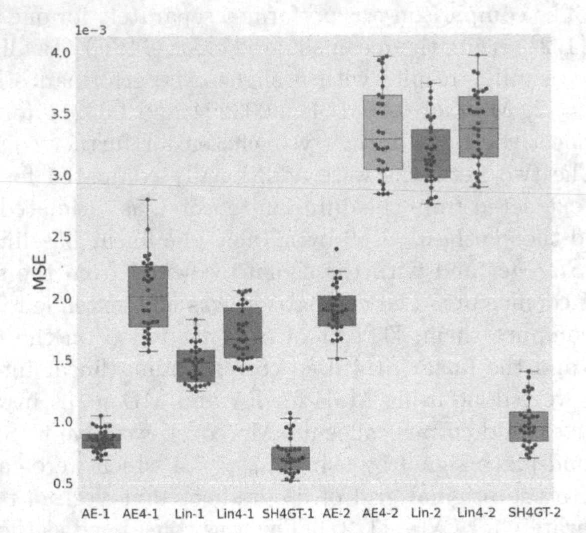

Fig. 2. Mean squared error (MSE) on dMRI images of autoencoder enhanced using *Avg-b1000-net* slices (AE-1, AE-2 for $N = 1, 2$ respectively) and for the baseline interpolation (linear on raw signal: Lin-1, Lin-2) and for *SH4-net* and SH linearly interpolated (Lin4-1, Lin4-2 for $N = 1, 2$ respectively). The lower bounds for the SH errors (SH4GT) were also included as a reference. (Method-1, Method-2 for synthesizing/interpolating $N = 1$ and $N = 2$ slices, respectively)

Comparing raw and SH domain enhancement (Fig. 2), we first observe that independently of the method (autoencoder or linear), working directly on the raw signal outperforms working on SH and projecting back to signal. In fact, autoencoder *Avg-b1000-net* outperforms linear interpolation, and for $N = 1$ it is closely comparable to the SH encoding (SH4GT-1 in Fig. 2). While the SH autoencoder enhancement underperforms the classical SH linear interpolation for $N = 1$, *SH4-net* slightly outperforms linear-SH for $N = 2$. This gap between $N = 1$ and $N = 2$ for SH linear and autoencoder can be explained by the rich information that the autoencoder was exposed to in the training phase from similar images compared to the interpolation that has solely access to local information.

3.2 FA and MD in Newborns

Comparing DTI scalar maps (Fig. 3) for the same previous configurations (see Fig. 2), we notice that the autoencoder enhancement outperforms the linear interpolation in all brain regions (except MD for cortical gray matter when removing one slice, i.e. $N = 1$) regardless of whether raw signal or SH was used. This outperformance is significant (paired Wilcoxon signed-rank test) for FA in all SH configurations, and for MD in one third of all configurations. The difference is typically more pronounced when we remove two slices ($N = 2$). Let us note that, opposite of what we observed at the DWI signal level, *SH4-net* outperforms linearly interpolated SH. Furthermore, for the FA map, *SH4-net* obtains the lowest mean squared errors, thus it is more suitable than autoencoder *Avg-b1000-net* or the linear interpolation. The opposite trend, i.e. *Avg-b1000-net* outperforming *SH4-net* with statistical significance, can be noticed for MD, with exception of the corpus callosum.

3.3 Qualitative Results of FA and MD in Fetuses

The DWIs synthesized by *SH4-net* using the latent space were visually consistent as they smoothly vary between the adjacent slices. Figure 4 displays the corresponding FA and MD maps for four subjects. We can clearly delineate the smooth transition between the two adjacent slices, especially in late gestational weeks fetuses in which the structures are more visible. For instance, the corpus callosum and the internal capsules of the synthesized slices displayed in FA maps are coherent with respect to their neighbouring slices.

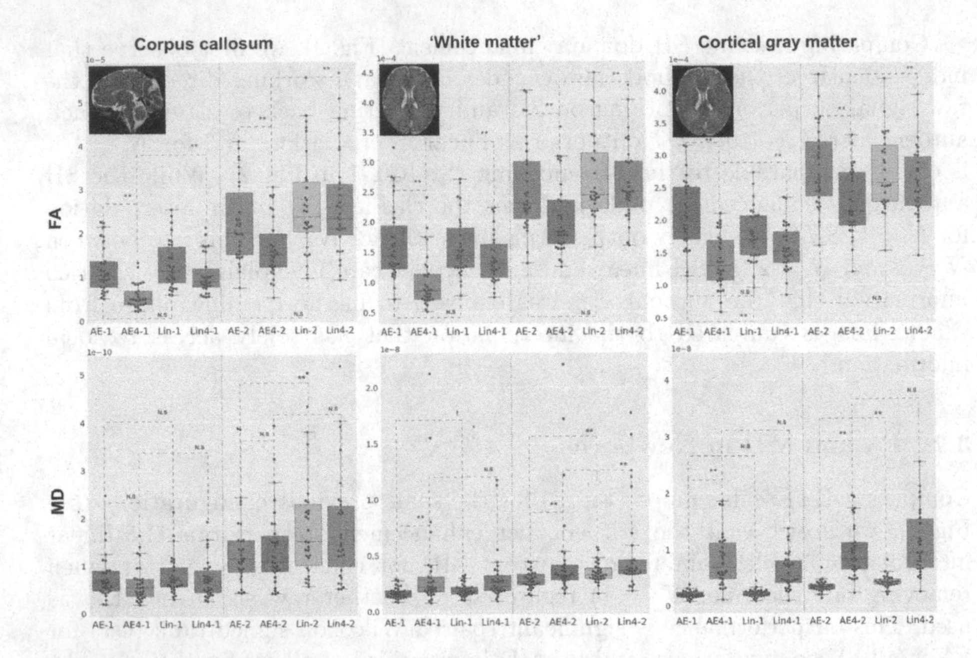

Fig. 3. Mean squared error of fractional anisotropy (FA) and mean diffusivity (MD) for different methods in three brain regions. See caption Fig. 2 for methods description. (Paired Wilcoxon signed-rank test: **: significant, p<0.028 - t: trending, p = 0.06 - N.S.: non significant: p>0.06)

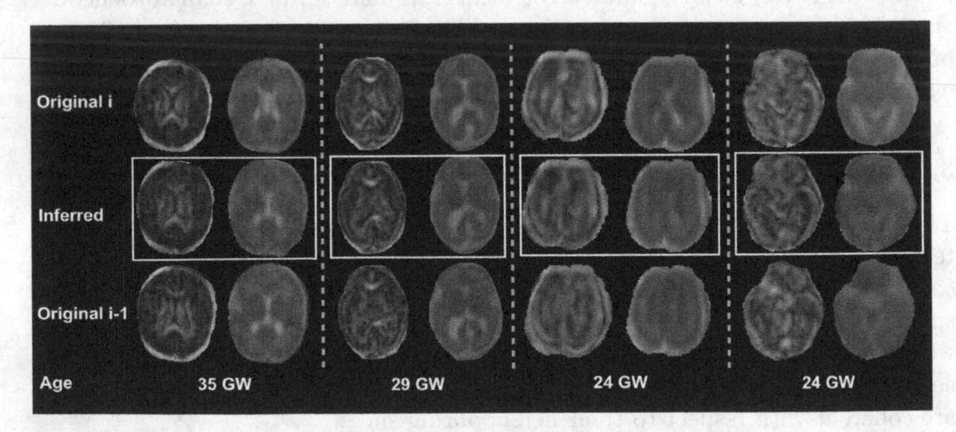

Fig. 4. Fractional anisotropy (FA) and mean diffusivity (MD) for four fetal subjects of respectively, from left to right, 4, 5, 4 and 4 mm of slice thickness. The middle row (red frames) illustrates synthesized slices corresponding to the diffusion tensor reconstructed with inferred DWI volumes with *SH4-net* and b0 with *b0-net*, using the two neighboring original slices (top and bottom rows). (Color figure online)

4 Conclusion

We have proposed autoencoders for dMRI through-plane slice inference in early brain development. The assessment was performed in both raw signal and spherical harmonics (SH) domains, where the latter proved to be more accurate for DTI-FA maps reconstruction and the former for raw data estimation. We hypothesize that this could be explained by some global bias introduced to the back projected raw signal by the SH trained autoencoder. However, the orientation information (i.e., signal's shape) was better preserved and hence, FA which is scale invariant, was clearly better depicted by SH autoencoder estimation. Lastly, we have successfully applied our method trained on newborn data to enhance the through-plane resolution of fetal data acquired in a different scanner, with a lower b-value and fewer gradient directions. Inferring missing slices or realistically increasing the through-plane resolution has to potential to translate to more accurate diffusion properties and hence a better uncovering of the underlying brain structure. In future work, we aim to increase the angular resolution in fetal images by using *supervised learning* to map spherical harmonics coefficients of order 4 (i.e., the maximal order that can be fit with clinical fetal images) to higher orders (6 or 8) using pre-term data.

Acknowledgments. This work was supported by the Swiss National Science Foundation (project 205321-182602, grant No. 185897: NCCR-SYNAPSY- "The synaptic bases of mental diseases" and the Ambizione grant PZ00P2_185814). We acknowledge access to the facilities and expertise of the CIBM Center for Biomedical Imaging, a Swiss research center of excellence founded and supported by Lausanne University Hospital (CHUV), University of Lausanne (UNIL), École polytechnique fédérale de Lausanne (EPFL), University of Geneva (UNIGE), Geneva University Hospitals (HUG) and the Leenaards and Jeantet Foundations.

References

1. Avants, B.B., Tustison, N., Song, G., et al.: Advanced normalization tools (ants). Insight J. **2**(365), 1–35 (2009)
2. Basser, P.J., Mattiello, J., LeBihan, D.: Mr diffusion tensor spectroscopy and imaging. Biophys. J . **66**(1), 259–267 (1994)
3. Bastiani, M., et al.: Automated processing pipeline for neonatal diffusion mri in the developing human connectome project. Neuroimage **185**, 750–763 (2019)
4. Batalle, D., Edwards, A.D., O'Muircheartaigh, J.: Annual research review: not just a small adult brain: understanding later neurodevelopment through imaging the neonatal brain. J. Child Psychol. Psychiatry **59**(4), 350–371 (2018)
5. Blumberg, S.B., Tanno, R., Kokkinos, I., Alexander, D.C.: Deeper image quality transfer: training low-memory neural networks for 3D images. In: Frangi, A.F., Schnabel, J.A., Davatzikos, C., Alberola-López, C., Fichtinger, G. (eds.) MICCAI 2018. LNCS, vol. 11070, pp. 118–125. Springer, Cham (2018). https://doi.org/10.1007/978-3-030-00928-1_14
6. Canales-Rodríguez, E.J., et al.: Sparse wars: a survey and comparative study of spherical deconvolution algorithms for diffusion MRI. Neuroimage **184**, 140–160 (2019)

7. Chatterjee, S., et al.: Shuffleunet: super resolution of diffusion-weighted mris using deep learning. arXiv preprint arXiv:2102.12898 (2021)
8. Chollet, F., et al.: keras (2015)
9. Clevert, D.A., Unterthiner, T., Hochreiter, S.: Fast and accurate deep network learning by exponential linear units (elus). arXiv preprint arXiv:1511.07289 (2015)
10. Dubois, J., Alison, M., Counsell, S.J., Hertz-Pannier, L., Hüppi, P.S., et al.: Mri of the neonatal brain: a review of methodological challenges and neuroscientific advances. J. Magn. Reson. Imaging **53**(5), 1318–1343 (2021)
11. Dyrby, T.B., et al.: Interpolation of diffusion weighted imaging datasets. Neuroimage **103**, 202–213 (2014)
12. Elsaid, N.M., Wu, Y.C.: Super-resolution diffusion tensor imaging using srcnn: a feasibility study. In: 2019 41st Annual International Conference of the IEEE Engineering in Medicine and Biology Society (EMBC), pp. 2830–2834. IEEE (2019)
13. Frank, L.R.: Characterization of anisotropy in high angular resolution diffusion-weighted mri. Magnetic Resonance in Med. Official J. Int. Soc. Magnetic Resonance Med. **47**(6), 1083–1099 (2002)
14. Garyfallidis, E., et al.: Dipy, a library for the analysis of diffusion mri data. Frontiers in neuroinformatics 8 (2014)
15. Hess, C.P., Mukherjee, P., Han, E.T., Xu, D., Vigneron, D.B.: Q-ball reconstruction of multimodal fiber orientations using the spherical harmonic basis. Magnetic Resonance Med. Official J. Int. Soc. Magnetic Resonance Med. **56**(1), 104–117 (2006)
16. Hüppi, P.S., Dubois, J.: Diffusion tensor imaging of brain development. In: Seminars in Fetal and Neonatal Medicine. vol. 11, pp. 489–497. Elsevier (2006)
17. Hutter, J., et al.: Time-efficient and flexible design of optimized multishell hardi diffusion. Magn. Reson. Med. **79**(3), 1276–1292 (2018)
18. Jakab, A., et al.: Disrupted developmental organization of the structural connectome in fetuses with corpus callosum agenesis. Neuroimage 111 (2015)
19. Jakab, A., Tuura, R., Kellenberger, C., Scheer, I.: In utero diffusion tensor imaging of the fetal brain: a reproducibility study. NeuroImage: Clinical 15 (2017)
20. Jha, R.R., Nigam, A., Bhavsar, A., Pathak, S.K., et al.: Multi-shell d-mri reconstruction via residual learning utilizing encoder-decoder network with attention (msr-net). In: 2020 42nd Annual International Conference of the IEEE Engineering in Medicine & Biology Society (EMBC), pp. 1709–1713. IEEE (2020)
21. Kebiri, H., et al.: Through-plane super-resolution with autoencoders in diffusion magnetic resonance imaging of the developing human brain. Frontiers in Neurology 13 (2022)
22. Kimpton, J., et al.: Diffusion magnetic resonance imaging assessment of regional white matter maturation in preterm neonates. Neuroradiology **63**(4), 573–583 (2021)
23. Kingma, D.P., Ba, J.: Adam: a method for stochastic optimization. arXiv preprint arXiv:1412.6980 (2014)
24. Koppers, S., Haarburger, C., Merhof, D.: Diffusion MRI signal augmentation: from single shell to multi shell with deep learning. In: Fuster, A., Ghosh, A., Kaden, E., Rathi, Y., Reisert, M. (eds.) MICCAI 2016. MV, pp. 61–70. Springer, Cham (2017). https://doi.org/10.1007/978-3-319-54130-3_5
25. Kuklisova-Murgasova, M., Estrin, G.L., Nunes, R.G., Malik, S.J., Rutherford, M.A., et al.: Distortion correction in fetal epi using non-rigid registration with a laplacian constraint. IEEE Trans. Med. Imaging **37**(1) (2017)
26. Makropoulos, A., et al.: Automatic whole brain mri segmentation of the developing neonatal brain. IEEE Trans. Med. Imaging **33**(9), 1818–1831 (2014)

27. Ning, L., et al.: A joint compressed-sensing and super-resolution approach for very high-resolution diffusion imaging. Neuroimage **125**, 386–400 (2016)
28. Ouyang, M., Dubois, J., Yu, Q., Mukherjee, P., Huang, H.: Delineation of early brain development from fetuses to infants with diffusion mri and beyond. Neuroimage **185**, 836–850 (2019)
29. Ramos-Llordén, G., et al.: High-fidelity, accelerated whole-brain submillimeter in vivo diffusion mri using gslider-spherical ridgelets (gslider-sr). Magn. Reson. Med. **84**(4), 1781–1795 (2020)
30. Sander, J., de Vos, B.D., Išgum, I.: Unsupervised super-resolution: creating high-resolution medical images from low-resolution anisotropic examples. In: Medical Imaging 2021: Image Processing, vol. 11596, p. 115960E. International Society for Optics and Photonics (2021)
31. Tournier, J.D., Calamante, F., Connelly, A.: Mrtrix: diffusion tractography in crossing fiber regions. Int. J. Imaging Syst. Technol. **22**(1), 53–66 (2012)
32. Tournier, J.D., Yeh, C.H., Calamante, F., Cho, K.H., Connelly, A., Lin, C.P.: Resolving crossing fibres using constrained spherical deconvolution: validation using diffusion-weighted imaging phantom data. Neuroimage **42**(2), 617–625 (2008)
33. Tustison, N.J., Avants, B.B., Cook, P.A., Zheng, Y., et al.: N4itk: improved n3 bias correction. IEEE Trans. Med. Imaging **29**(6), 1310–1320 (2010)
34. Veraart, J., Fieremans, E., Novikov, D.S.: Diffusion mri noise mapping using random matrix theory. Magn. Reson. Med. **76**(5), 1582–1593 (2016)

Super-Resolution of Manifold-Valued Diffusion MRI Refined by Multi-modal Imaging

Tyler A. Spears(✉) and P. Thomas Fletcher

Department of Electrical and Computer Engineering, University of Virginia, Charlottesville, VA, USA
{tas6hh,ptf8v}@virginia.edu

Abstract. Diffusion MRI is the foundation for understanding the structures and disorders in the human connectome, but low spatial resolution fundamentally limits this understanding. Methods for increasing DTI resolution *post hoc* must carefully utilize all available information to reduce bias and uncertainty in this ill-conditioned inverse problem. Previous machine learning approaches have largely cast this problem as a standard super-resolution task without taking advantage of domain knowledge surrounding DTIs. Outside this domain, recent work in super-resolution has given attention to preserving an input's fine-scale information as it is passed through a network. Our contribution consists of a novel deep learning model for DTI super-resolution with three important advancements: 1) a novel procedure for refining DTI predictions with high-resolution T2-weighted images, 2) interpolation over log-Euclidean tensors that is immune to the "swelling effect," and 3) the effective use of densely-connected residual networks that preserve detail from the input. Through experiments on HCP data, we show that our model achieves the best performance in the literature for increasing the resolution of DTIs. We further analyze the effect of each proposed component with thorough model ablation tests.

Keywords: Super-resolution · Diffusion tensor imaging · Log euclidean

1 Introduction

Since its invention nearly 30 years ago, diffusion MRI (dMRI) has remained the only imaging modality capable of studying the connections in the human brain non-invasively and *in vivo*. This is made possible through methods that estimate local diffusion models and subsequently infer fiber orientations from the diffusion data. In both the clinic and the dMRI literature, the diffusion tensor image (DTI) model is the most widespread and well understood of such models [6]. While many other diffusion representations exist, such as those in diffusion spectrum imaging [21] and q-ball imaging [18], all suffer from low spatial resolution in

S. Cetin-Karayumak et al. (Eds.): CDMRI 2022, LNCS 13722, pp. 14–25, 2022.
https://doi.org/10.1007/978-3-031-21206-2_2

the diffusion-weighted images (DWIs) themselves. Compared to structural T1-weighted (T1w) or T2-weighted (T2w) MRIs, dMRIs are acquired at roughly half the spatial resolution. This poses serious limitations in dMRI applications. For example, clinicians planning a treatment based on a lesion's position relative to a fiber tract. Another example is tractography, where low resolution induces severe partial volume effects, confounding signals from overlapping fiber tracts. Consequently, tractography methods struggle to distinguish tracts when fibers are crossing, "kissing," or otherwise laid out in complex configurations.

Unfortunately, the MRI sequences needed for measuring diffusion fundamentally require a trade-off between spatial resolution, scan acquisition time, and signal-to-noise ratio. Consequently, methods of reconstructing high-resolution dMRIs have existed almost as long as dMRI itself. One of the first such methods relied on sub-pixel spatial shifts while acquiring low-resolution diffusion-weighted images (DWIs) [12]. A high-resolution DWI was estimated with iterative back-projection. Later methods treated super-resolution as a general reconstruction problem performed *post hoc*, reducing dependency on the scan sequence. For example, Nedjati-Gilani et al. [11] used damped least-squares with a spatial coherence constraint to recover fiber orientation based on neighborhood anisotropy.

Machine learning (ML) methods have largely replaced more traditional image processing algorithms due to increased performance and the availability of the high-quality dMRI data from the Human Connectome Project (HCP) [19]. The first of these was in Alexander et al. [3], where authors used random forest regression to learn a map between downsampled HCP images and their full resolution counterparts; this approach was named image quality transfer (IQT). Tanno et al. [16] improved on this with a simple 3-layer convolution (conv, for brevity) network. Blumberg et al. [7] extended their conv network with "reversible layers," which required less memory in backpropagation. Other works expanded the scope of single-image super-resolution (SISR) in diffusion. Qin et al. [13] added a secondary network that used DWIs and high-resolution T1w images to predict high resolution NODDI parameters. Anctil-Robitaille et al. [4] took the use of T1w images further and synthesized entire DTI volumes from only T1w images.

In this work, we propose an ML model that builds upon these works and produces accurate, high-resolution DTIs. Our novel contributions are driven by the observation that DTI SISR requires algorithms that preserve the information in the low-resolution input DTI throughout the model, while also taking advantage of information outside the input. In the following sections, we describe a novel deep learning model that effectively leverages information from low-resolution DTIs and high-resolution T2w images, while respecting the nature of DTIs being in the space of symmetric positive-definite (SPD) matrices.

2 Background and Methods

2.1 Background

Structural MRI with Diffusion. Structural MRIs have previously been used with diffusion to compensate for weaknesses inherent to dMRI. In Sect. 1, we gave specific examples such methods used with ML models [4,13]. Other works have treated T2w images as a low-distortion reference of the $b = 0$ DWIs to compensate for EPI distortions [17]. We note that the dependency on high-resolution T1w or T2w scans is reasonable, even in the clinic, as the (time-consuming) diffusion scans are usually accompanied by (relatively quick) structural scans.

Extracting and representing the information from T1w or T2w images must be handled carefully in SISR. With ML models, as discussed by Qin et al. [13], simple concatenation of DTIs and anatomical MRIs at a network's input would require manipulating the spatial scale of at least one volume. In general, combining multi-modal MRIs in such black-box models risks creating undesirable bias learned from features in the anatomical image. For example, a network could maintain low prediction error by mapping certain T2-weighted textures to their most common tensor component values, but such a mapping would perform poorly on subjects with conditions only visible in the diffusion maps.

DTIs on the SPD Manifold. The first and most widespread diffusion model in dMRI is the DTI [6], which is estimated at a given voxel as a 3×3 SPD matrix. Ideally, such matrices could be analyzed using standard Euclidean distances. However, processing DTIs with the standard Euclidean distance, e.g., in interpolation or filtering, leads to physically unrealistic tensors that suffer from a *swelling effect* [5,8], i.e., a decrease in anisotropy and increase in determinant.

To resolve this, Arsigny et al. [5] proposed the use of Riemannian metrics on the matrix logarithm of the DTIs, the log-Euclidean (LE) metrics. Given an SPD diffusion tensor $\mathbf{D}$, let $\mathbf{D} = \mathbf{U}\boldsymbol{\Lambda}\mathbf{U}^T$ be its eigendecomposition. Then let the matrix logarithm of $\mathbf{D}$, noted as $\mathbf{T}$, have the eigendecomposition of $\mathbf{T} = \mathbf{U}\tilde{\boldsymbol{\Lambda}}\mathbf{U}^T$. The mapping between $\mathbf{D}$ and $\mathbf{T}$ is given by

$$\mathbf{T} = \mathrm{Log}(\mathbf{D}) = \mathbf{U}\begin{bmatrix} \log\lambda_1 & & \\ & \log\lambda_2 & \\ & & \log\lambda_3 \end{bmatrix}\mathbf{U}^T$$

$$\mathbf{D} = \mathrm{Exp}(\mathbf{T}) = \mathbf{U}\begin{bmatrix} \exp\tilde{\lambda}_1 & & \\ & \exp\tilde{\lambda}_2 & \\ & & \exp\tilde{\lambda}_3 \end{bmatrix}\mathbf{U}^T$$

In this logarithmic domain, the Frobenius distance $\|\mathbf{T}_1 - \mathbf{T}_2\|_F$ is the geodesic distance under a particular Riemannian metric, and we use this distance in our objective function during optimization. Given that $\mathbf{T}$ is a symmetric matrix by the definition of $\mathbf{D}$, computation can be simplified by indexing and scaling the lower triangular values in $\mathbf{T}$ as a 6-vector such that $\|\mathbf{T}_i - \mathbf{T}_j\|_F$ is equivalent to the vector L_2 distance $\|\mathrm{V}_i - \mathrm{V}_j\|_2$. Here, for any given $\mathbf{T} = \mathbf{T}_i$,

$$\mathrm{V} = \left[\mathbf{T}_{1,1}\ \sqrt{2}\mathbf{T}_{2,1}\ \mathbf{T}_{2,2}\ \sqrt{2}\mathbf{T}_{3,1}\ \sqrt{2}\mathbf{T}_{3,2}\ \mathbf{T}_{3,3}\right]. \tag{1}$$

Super-Resolution with Deep Learning. Super-resolution methods from the computer vision literature often serve as starting points for medical image processing. Recently, works in SISR have developed architectures and transformations suited to the precise, computationally sensitive problem of artificial resolution enhancement. Two prominent developments are the efficient sub-pixel conv neural network (ESPCN) channel shuffling layer [14], and the use of densely-connected residual layers [1]. The ESPCN layer casts SISR as a prediction of offsets, rather than regressing to a higher-resolution image directly. This layer has been utilized in DTI resolution enhancement as in Blumberg et al. [7] and Tanno et al. [16]. ESPCN has been further improved to minimize checkerboard artifacts in the output. Such additions include a nearest-neighbor weight initialization of the ESPCN conv layer, known as initialization to conv network resize (ICNR) [2], and a small average pooling performed on the output [15].

Other recent SISR methods highlight the sensitivity of resolution enhancement to the loss of scale information throughout the network. Residual conv layers preserve information by only predicting an *additive map* over the input, which reduces the distortion a network can induce relative to performing a full conv layer [1]. Between-layer dense connections, also called cascading layers, provide "shortcuts" for information to be passed from earlier layers to those deeper in the network, relaxing the information dependencies between layers. Of note is the cascading residual network (CARN) [1], a high-performing, lightweight architecture that exemplifies information preservation through previously mentioned techniques and the removal of normalization layers entirely. The CARN model forms the backbone network behind our proposed method.

2.2 Proposed Method

Our proposed method is illustrated in Fig. 1. In essence, the method consists of a lightweight CARN-based network that operates on the Log-mapped DTIs. Given a subject's low-resolution DTI volume, $\mathbf{D}$, and their corresponding high-resolution T2w structural volume, $\mathbf{S}$, we wish to predict a high-resolution DTI, $\overset{*}{\mathbf{D}}$. We compute the Log map of each voxel in the input and target DTIs, $\mathbf{T} = \text{Log}(\mathbf{D})$ and $\overset{*}{\mathbf{T}} = \text{Log}(\overset{*}{\mathbf{D}})$, respectively. To lessen the amount of redundant computation, we reduce $\mathbf{T}$ from a volume of 3×3 matrices to a volume of 6-vectors $\mathrm{V} = s(\mathbf{T})$ (where $s(\cdot)$ follows Eq. 1) when processing inputs through the network. We use the matrix notation $\mathbf{T}$ and $\mathbf{D}$ for notational brevity.

The network $F(\cdot\,; \boldsymbol{\Theta})$, parameterized by $\boldsymbol{\Theta}$, is illustrated in Fig. 1. Our proposed network is a densely-connected conv network built up from basic residual blocks, similar to those found in CARN [1] and illustrated in Fig. 1A. The residual blocks are stacked into so-called cascade blocks (see Fig. 1B). These cascade blocks themselves are also connected in a cascading pattern, which forms the two sub-networks of our model. As shown in Fig. 1C, the first sub-network only takes the low-resolution $\mathbf{T}$ as input, where the sub-network has three cascade blocks. This model's output is given to an upsampling layer, which consists of a 1^3 ICNR-initialized conv layer, an ESPCN shuffling layer, and a final 2^3 average pooling layer padded by 1^3 to maintain the correct upsampled shape [2,15]. The upsampling layer increases the resolution by $2\times$ in each spatial dimension. This output is copied, with one copy being passed off as a "pre-anatomical" prediction $\tilde{\mathbf{T}}_{\text{pre}}$, and another passed to the anatomical refinement sub-network. The structural volume $\mathbf{S}$ is concatenated and merged channel-wise with the upsample layer's output. The refinement sub-network processes this merged volume

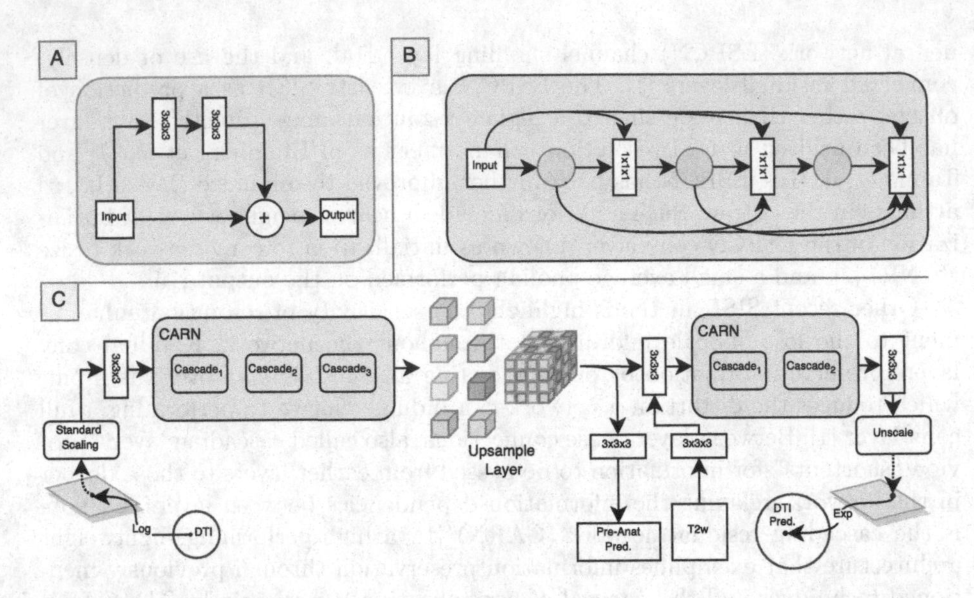

Fig. 1. The proposed model for SISR on LE matrices. A) A basic residual block with two conv layers that learns an additive transformation of the input. B) A densely-connected cascade block. Residual blocks (red circles, shown in A) and 1^3 conv layers are stacked, with layer outputs "cascading" to subsequent layers. C) The full proposed model with two CARN sub-networks; blue rectangles are cascade blocks as shown in B. The blue rectangle containing blue rectangles indicates a "cascade of cascades", where each cascade module itself is connected as shown in B. This is the CARN module. The upsample layer is a modified ESPCN layer (see Sect. 2.2). (Color figure online)

through a CARN model with two cascade blocks. The final network output is mapped to Euclidean space with Exp, and the final prediction is given as $\tilde{\mathbf{T}}$.

To approach the problem of multi-modal image fusion as laid out in Sect. 2.1, we promote a soft "compartmentalization" of the sub-networks. The first sub-network maximizes its prediction based solely upon the low-resolution diffusion tensor, and the second sub-network serves as a minimal "refinement" step given a T2w image $\mathbf{S}$. At minimum, our proposed model can be inspected through its pre-anatomical predicted DTI $\tilde{\mathbf{D}}_{\text{pre}} = \text{Exp}(\tilde{\mathbf{T}}_{\text{pre}})$. We further promote this separation by calculating training loss over both the output of the ESPCN upsampling layer and the final output prediction:

$$\ell = \lambda \|\tilde{\mathbf{T}}_{\text{pre}} - \overset{*}{\mathbf{T}}\|_F + (1 - \lambda)\, \|\tilde{\mathbf{T}} - \overset{*}{\mathbf{T}}\|_F, \tag{2}$$

with the Frobenius norm $\|\cdot\|_F$, ground truth $\overset{*}{\mathbf{T}}$, predictions $\tilde{\mathbf{T}}_{\text{pre}}$ and $\tilde{\mathbf{T}}$, and regularization term λ.

3 Experiments

3.1 Data

All data were sampled from the HCP Young Adult Study [19], with all the acquisition parameters and HCP pre-processing methods described in Glasser et al. [9]. We randomly selected 48 subjects with dMRI scans for the training, validating, and testing steps in our experiments. All DWIs were limited to scans with $b \leq 1600\,\mathrm{s/mm}^2$. The high-resolution ground truth data were processed in the original resolution of $1.25\,\mathrm{mm}^3$, while the low-resolution data were downsampled by a factor of 2 in the DWIs with a mean filter of size 2. The full-resolution and low-resolution DWIs were independently fitted to their own DTIs using weighted least-squares regression. Each subject's T2w image was downsampled from 0.7 mm isotropic to 1.25 mm isotropic with cubic spline interpolation.

During our experiments, we found that the HCP DTIs contained a small number of outlier voxels located at barriers between the brain and skull. These outliers were orders of magnitude larger than was biologically possible, resulting in less stable training and biased quantitative metrics. So, as a preprocessing step, we clipped the eigenvalues of all subjects to the range of $[1.0 \times 10^{-5},\ 3.32008 \times 10^{-3}]\,\mathrm{mm/s}^2$. This upper bound was found by taking the eigenvalues of subjects' cerebral spinal fluid, which should contain the highest diffusivity values in the brain, and calculating the 99th percentile of the eigenvalue distribution. This resulted in more consistent model training and testing, with a negligible effect on the DTIs themselves. However, we found that this clipping *improved* scores for all models, including RevNet4, compared to past IQT reports, [7,16]. So, our quantitative results cannot be directly compared to these works.

3.2 Implementation and Training Details

During training, subjects were randomly assigned to different subsets, with 10 for training, 4 for validation, and 34 for testing. Similar to Tanno et al. [16] and Blumberg et al. [7], we increased sample heterogeneity during training by using only small regions of subject DTIs. For every training epoch, we sampled 4,000 random patches of size 24^3 for every training subject; each subject had patches re-sampled at the start of every epoch. These patches were selected such that every patch had at least one voxel found in the subject's brain mask (provided in the HCP dataset). Both the input and target were sampled to produce a pair of patches at the same location in the subject's data. These patch pairs were collated into batches of size 32, and the model was trained for 50 epochs.

The model was optimized with the Adam with weight decay (AdamW) algorithm [10], with a rate of 2.5×10^{-4}, $\beta_1 = 0.9$, $\beta_2 = 0.999$, and weight decay $\lambda_{\text{adamw}} = 0.01$. The activation function was the exponential linear unit (ELU). For training, the loss regularization term was $\lambda = 0.35$ (Eq. 2). This was selected such that the ratio of $\tilde{\mathbf{T}}$ loss to $\tilde{\mathbf{T}}_{\text{pre}}$ loss was approximately 2:1. In each backwards pass, the L_2 norms of the gradients were clipped to a maximum of 0.25 to help training stability. Models were implemented in the Pytorch library, version 1.10.2. To reduce memory usage and lower training time, network parameters were cast to half-precision (16-bit) floating point values. Input and ground truth DTIs were scaled such that the foreground voxels of each subject and each tensor component followed a standard normal distribution; the same was performed on the T2w inputs. During validation and testing, the model predictions and ground truths were un-scaled before their similarity was calculated. For other implementation details, see our github repository and results files publicly available online[1].

3.3 Models and Evaluation Metrics

Our primary experiments were run over three different models. As a baseline, cubic spline interpolation was performed on both the DTIs and the matrix logarithms. For comparison to a previous work, we trained and tested the RevNet4 [7] for 50 epochs. RevNet4 was trained with outliers removed as described in Sect. 3.1, patches re-sampled every epoch (otherwise leading to overfitting), the mean Frobenius distance as its loss function, and normal backpropagation performed without output caching; other training parameters are as described in Blumberg et al. [7]. Finally, we trained our proposed model on log-Euclidean tensors.

We ran further ablation experiments to analyze three different effects: 1) training on Euclidean DTIs vs. LE tensors, 2) including high-resolution T2w inputs (for those models with the refinement sub-network), and 3) increasing the number of network weights with the refinement sub-network. Models trained on Euclidean DTIs (the "DTI" models) were trained as described in Sects. 2.2 and 3.2, but without the Log and Exp mappings. Models without an anatomical refinement sub-network are shown as "No Anat Net." Models with this sub-network that were not given anatomical images (the "Fake Anat" models) replaced those inputs with an equally-sized **1**-tensor (the "fake" T2w).

We report four metrics in total: 1) square root of the mean Euclidean distance on DTIs (DTI RMFD), 2) square root of the mean Frobenius distance on log-Euclidean tensors (LE RMFD), 3) peak signal-to-noise ratio (PSNR) of the DTIs where tensor components are considered as "channels", and 4) the mean structural similarity index [20] of the fractional anisotropy (SSIM FA). We compute SSIM over FA because SSIM is based on *human perception.* DTIs are rarely viewed directly by clinicians and are not meant for direct human interpretation, while FA maps are very commonly interpreted for clinical purposes. Each model prediction was made in the model's training domain, then Log or Exp mapped to or from the domain of the metric. For example, models trained on LE tensors have their outputs Exp mapped back into DTIs for measuring the DTI RMFD. Cubic spline interpolation was calculated both on the DTIs and the LE tensors.

[1] https://osf.io/r37v5.

Table 1. Average performance scores over test sets for all models, ± standard deviation. Best average scores are printed in bold. Each sample in a model's performance distribution is the mean metric over a subject's image in the test set, over each train-test split; 34 × 3 samples in total, for each model and metric. The cubic spline distribution was estimated from the average score over each subject.

model \ metric	DTI RMFD ↓ $\times 10^{-2}$	LE RMFD ↓	PSNR ↑	FA SSIM ↑
Cubic Spline	0.050 ± 0.003	1.055 ± 0.030	43.746 ± 0.107	0.881 ± 0.010
RevNet4	0.038 ± 0.001	0.825 ± 0.055	44.609 ± 0.121	0.883 ± 0.011
CARN LE (Proposed)	0.036 ± 0.003	$\mathbf{0.684 \pm 0.047}$	44.595 ± 0.182	$\mathbf{0.919 \pm 0.008}$
CARN DTI	$\mathbf{0.033 \pm 0.003}$	0.764 ± 0.060	$\mathbf{44.778 \pm 0.190}$	0.916 ± 0.009
CARN LE No Anat Net	0.042 ± 0.002	0.724 ± 0.041	44.450 ± 0.129	0.907 ± 0.008
CARN DTI No Anat Net	0.038 ± 0.001	0.820 ± 0.055	44.578 ± 0.125	0.899 ± 0.009
CARN LE Fake Anat	0.041 ± 0.002	0.722 ± 0.042	44.527 ± 0.123	0.909 ± 0.008
CARN DTI Fake Anat	0.037 ± 0.001	0.820 ± 0.056	44.611 ± 0.127	0.899 ± 0.009

4 Results

4.1 Proposed Model Performance

We provide all model performance scores in Table 1. To show the variance in model results, Fig. 3 illustrates the distribution of test subject performance for each model. In every performance metric, our proposed CARN model, either trained on LE tensors or Euclidean DTIs, matches or exceeds the performance of RevNet4, on average. When looking at the distribution of model performances on subjects in the different test sets, RevNet4 only comes within one standard deviation of the full CARN model on the PSNR metric.

Qualitatively, CARN LE predicts overall improved DTIs compared to RevNet4. We use the examples in Fig. 2. Starting with the colored fiber direction map, the proposed model accurately resolves edges between regions with high and low FA and preserves voxels containing thin "strands" of oriented fibers. This is most evident in the thin white matter forming the outer-most gyri of the brain.

However, FA is derived from equally-weighted differences in all the DTI's eigenvalues, and may "average out" some prediction errors. Looking at the $D_{x,x}$ on-diagonal component of the diffusion tensor, we see even more advantages for the proposed CARN model. High-frequency details and edges are more evident in the CARN LE model, whereas edges in the cortical gyri are almost completely lost in RevNet4. Some details found in the predicted $D_{x,x}$ component can be seen as cerebral spinal fluid regions found directly in the T2w input. However, the proposed model also flexibly incorporates information for all tensor components, as seen in the $D_{y,z}$ comparison. Here, our proposed model provides a distinct edge near the corpus callosum, even in components with relatively low contrast.

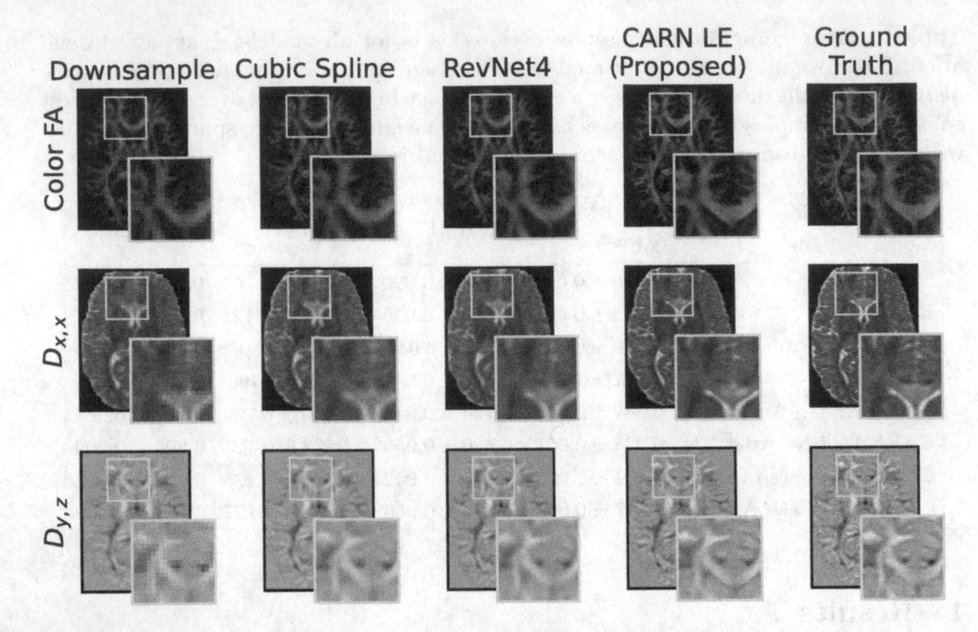

Fig. 2. Visual comparison of predictions between models. Each column is a model's prediction, and each row is a DTI measurement. A zoomed-in region of interest is outlined in yellow. All images are from predictions of the same subject in the test set (never in the training set) at the same axial slice. Color FA: FA-weighted color direction map; $D_{x,x}$: on-diagonal component of the diffusion tensor; $D_{y,z}$: off-diagonal component of the diffusion tensor. (Color figure online)

4.2 Model Ablation Results

The performance results for our model ablation experiments are quantified in Table 1, and the full distributions of each model's test performances are given in Fig. 3. To test the effect of adding the refinement sub-network without any real T2w input, we compare the "No Anat Net" models to the "Fake Anat" models. Consistently, there is little to no difference across all metrics when marginalizing out model training domain, indicating that the additional parameters given by the sub-network provide very little improvement on their own. Thus, we can examine other effects with minimal concern for the confound of model size.

We also analysed the effect of training the models on LE tensors vs. DTIs. Here, when the model definition is marginalized out, performance differences are primarily affected by the metric used (the test domain). Models trained on DTIs perform better on DTI RMFD and PSNR, while models trained on LE tensors perform better on LE RMFD and SSIM FA. For both RMFD metrics, this performance gap is explained by the use of RMFD in the objective function itself - DTI models were trained to minimize DTI RMFD, and similar for LE models and LE RMFD. PSNR is also based on the mean-squared error of the DTIs, which is proportional to the DTI RMFD, giving a similar advantage to DTI networks. Explaining the advantage of LE models in FA SSIM is less obvious.

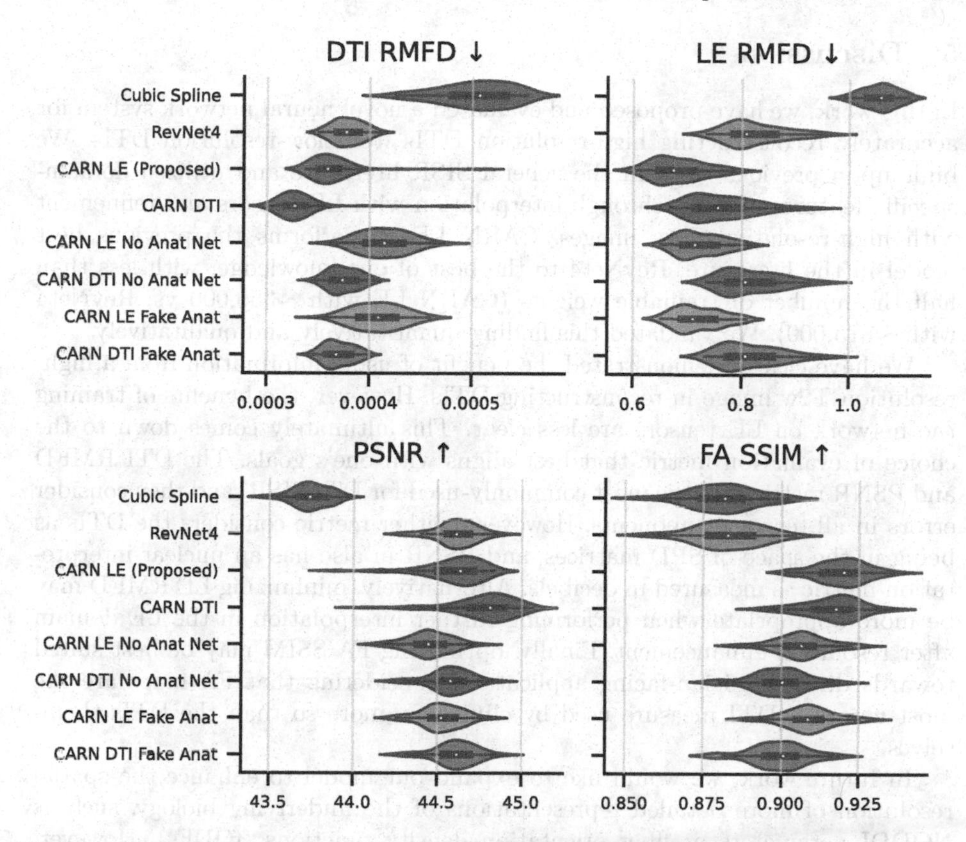

Fig. 3. Test performance distributions of all models tested shown as violin plots with nested box plots. Each subplot is a separate performance metric. See Table 1 for a description of the model performance distributions.

The score improvements are modest, especially in the full proposed model (with anatomical refinement). One possible cause is that each LE model guarantees that all diffusion tensor eigenvalues will be greater than 0, while the DTI models have no such guarantee. In practice, this is an uncommon issue for all CARN models, but its rarity may explain the modest difference in performance.

Finally, the benefits of anatomical refinement are shown directly when comparing models without T2w images ("Fake Anat" models) to the full proposed CARN models (CARN LE and CARN DTI). In every metric, either CARN DTI or CARN LE achieved the best performance out of all other tested models. However, this increase in performance is accompanied by a slight increase in performance variance. Some of this variance may be explained by a small number of outlier subjects (for example, DTI RMFD scores shown in Fig. 3). We leave the characterization of this sensitivity as a question for future research.

5 Discussion

In this work, we have proposed and evaluated a novel neural network system for accurately reconstructing high-resolution DTIs from low-resolution DTIs. We built upon previous work in the general SISR literature and utilized domain-specific features of DTIs through interpolation with LE tensors and refinement with high-resolution T2w images. CARN LE outperforms the previous best model in the literature, RevNet4 to the best of our knowledge, with less than half the number of trainable weights (CARN LE with ∼350,000 vs. RevNet4 with ∼875,000). We validated this finding quantitatively and qualitatively.

We have clearly demonstrated the benefit of using information from a high-resolution T2w image in reconstructing DTIs. However, the benefits of training the network on LE tensors are less clear. This ultimately comes down to the choice of evaluation metric that best aligns with one's goals. The DTI RMFD and PSNR metrics are the most commonly-used for DTI SISR, and they consider errors in all tensor components. However, neither metric considers the DTIs as being in the space of SPD matrices, and PSNR in also has an unclear interpretation metric as measured in decibels. Alternatively, minimizing LE RMFD may be more appropriate when performing further interpolation in the LE domain after resolution enhancement. Finally, optimizing FA SSIM may be best suited towards direct clinician-facing applications considering that FA maps are the most common DTI measure used by clinicians, more so than the DTIs themselves.

In future work, we would like to expand our model to enhance the spatial resolution of more detailed representations of the underlying biology, such as NODDI parameters or fiber orientation density functions (fODFs). However, both estimations would require as input all DWIs of all gradient strengths and directions that are not consistent between datasets, requiring a more flexible signal parameterisation than the 6-parameter DTI [13]. We would also like to extend our model towards improving interpolation performed in many tractography algorithms. The possibility of highly detailed, precise, and accurate maps of an individual patient's connectome, available to the average clinician, would fulfill one of the most important promises made by diffusion MRI.

Acknowledgments. This research was supported by the UVA Brain Institute Presidential Fellowship in Neuroscience. Data were provided [in part] by the Human Connectome Project, WU-Minn Consortium (Principal Investigators: David Van Essen and Kamil Ugurbil; 1U54MH091657) funded by the 16 NIH Institutes and Centers that support the NIH Blueprint for Neuroscience Research; and by the McDonnell Center for Systems Neuroscience at Washington University.

References

1. Ahn, N., Kang, B., Sohn, K.-A.: Fast, accurate, and lightweight super-resolution with cascading residual network. In: Ferrari, V., Hebert, M., Sminchisescu, C., Weiss, Y. (eds.) ECCV 2018. LNCS, vol. 11214, pp. 256–272. Springer, Cham (2018). https://doi.org/10.1007/978-3-030-01249-6_16

2. Aitken, A.P., Ledig, C., Theis, L., Caballero, J., Wang, Z., Shi, W.: Checkerboard artifact free sub-pixel convolution: a note on sub-pixel convolution, resize convolution and convolution resize. ArXiv abs/1707.02937v1 [cs.CV] (2017)
3. Alexander, D.C., Zikic, D., Zhang, J., Zhang, H., Criminisi, A.: Image quality transfer via random forest regression: applications in diffusion mri. In: MICCAI 2014, pp. 225–232 (2014)
4. Anctil-Robitaille, B., Desrosiers, C., Lombaert, H.: Manifold-aware CycleGAN for high-resolution structural-to-dti synthesis. In: CDMRI 2021, pp. 213–224 (2021)
5. Arsigny, V., Fillard, P., Pennec, X., Ayache, N.: Log-Euclidean metrics for fast and simple calculus on diffusion tensors. Magn. Reson. Med. **56**(2), 411–421 (2006)
6. Basser, P.J., Mattiello, J., Lebihan, D.: Estimation of the effective self-diffusion tensor from the NMR Spin Echo. J. Magn. Res., Series B **103**(3), 247–254 (1994)
7. Blumberg, S.B., Tanno, R., Kokkinos, I., Alexander, D.C.: Deeper image quality transfer: training low-memory neural networks for 3d images. In: MICCAI 2018, pp. 118–125 (2018)
8. Fletcher, P.T., Joshi, S.: Principal geodesic analysis on symmetric spaces: statistics of diffusion tensors. In: CVAMIA 2004, pp. 87–98 (2004)
9. Glasser, M.F., Sotiropoulos, S.N., Wilson, J.A., Coalson, T.S., Fischl, B., Andersson, J.L., Xu, J., Jbabdi, S., Webster, M., Polimeni, J.R., Van Essen, D.C., Jenkinson, M.: The minimal preprocessing pipelines for the Human Connectome Project. Neuroimage **80**, 105–124 (2013)
10. Loshchilov, I., Hutter, F.: Fixing weight decay regularization in Adam. ArXiv abs/1711.05101v3 [cs.LG] (2017)
11. Nedjati-Gilani, S., Alexander, D.C., Parker, G.J.M.: Regularized super-resolution for diffusion MRI. In: ISBI 2008, pp. 875–878 (2008)
12. Peled, S., Yeshurun, Y.: Superresolution in MRI: application to human white matter fiber tract visualization by diffusion tensor imaging. Magn. Reson. Med. **45**(1), 29–35 (2001)
13. Qin, Y., Li, Y., Zhuo, Z., Liu, Z., Liu, Y., Ye, C.: Multimodal super-resolved q-space deep learning. Med. Im. Analysis, p. 102085, April 2021
14. Shi, W., et al.: Real-time single image and video super-resolution using an Efficient Sub-Pixel Convolutional Neural Network. In: CVPR 2016, pp. 1874–1883 (2016)
15. Sugawara, Y., Shiota, S., Kiya, H.: Super-resolution using convolutional neural networks without any checkerboard artifacts. In: ICIP 2018, pp. 66–70 (2018)
16. Tanno, R., Worrall, D., Kaden, E., Ghosh, A., Grussu, F., Bizzi, A., Sotiropoulos, S., Criminisi, A., Alexander, D.: Uncertainty modelling in deep learning for safer neuroimage enhancement: demonstration in diffusion MRI. Neuroimage **225**, 117366 (2021)
17. Tao, R., Fletcher, P.T., Gerber, S., Whitaker, R.T.: A variational image-based approach to the correction of susceptibility artifacts in the alignment of diffusion weighted and structural MRI. In: IPMI 2009, vol. 21, pp. 664–675 (2009)
18. Tuch, D.S.: Q-ball imaging. Magn. Reson. Med. **52**(6), 1358–1372 (2004)
19. Van Essen, D.C., Smith, S.M., Barch, D.M., Behrens, T.E.J., Yacoub, E., Ugurbil, K.: The WU-Minn human connectome project: an overview. Neuroimage **80**, 62–79 (2013)
20. Wang, Z., Bovik, A., Sheikh, H., Simoncelli, E.: Image quality assessment: from error visibility to structural similarity. IEEE TIP **13**(4), 600–612 (2004)
21. Wedeen, V.J., Hagmann, P., Tseng, W.Y.I., Reese, T.G., Weisskoff, R.M.: Mapping complex tissue architecture with diffusion spectrum magnetic resonance imaging. Magn. Reson. Med. **54**(6), 1377–1386 (2005)

Lossy Compression of Multidimensional Medical Images Using Sinusoidal Activation Networks: An Evaluation Study

Matteo Mancini[1], Derek K. Jones[1], and Marco Palombo[1,2](✉)

[1] Cardiff University Brain Research Imaging Centre (CUBRIC), Cardiff University, Cardiff, UK
{mancinim,jonesd27,palombom}@cardiff.ac.uk
[2] School of Computer Science and Informatics, Cardiff University, Cardiff, UK

Abstract. In this work, we evaluate how neural networks with periodic activation functions can be leveraged to reliably compress large multidimensional medical image datasets, with proof-of-concept application to 4D diffusion-weighted MRI (dMRI). In the medical imaging landscape, multidimensional MRI is a key area of research for developing biomarkers that are both sensitive and specific to the underlying tissue microstructure. However, the high-dimensional nature of these data poses a challenge in terms of both storage and sharing capabilities and associated costs, requiring appropriate algorithms able to represent the information in a low-dimensional space. Recent theoretical developments in deep learning have shown how periodic activation functions are a powerful tool for implicit neural representation of images and can be used for compression of 2D images. Here we extend this approach to 4D images and show how any given 4D dMRI dataset can be accurately represented through the parameters of a sinusoidal activation network, achieving a data compression rate about 10 times higher than the standard DEFLATE algorithm. Our results show that the proposed approach outperforms benchmark ReLU and Tanh activation perceptron architectures in terms of mean squared error, peak signal-to-noise ratio and structural similarity index. Subsequent analyses using the tensor and spherical harmonics representations demonstrate that the proposed lossy compression reproduces accurately the characteristics of the original data, leading to relative errors about 5 to 10 times lower than the benchmark JPEG2000 lossy compression and similar to standard pre-processing steps such as MP-PCA denosing, suggesting a loss of information within the currently accepted levels for clinical application.

Keywords: Data compression · Multidimensional imaging · Diffusion-weighted MRI · Deep learning · Neural implicit representation

1 Introduction

Modern medical imaging provides both structural and functional information on anatomical features and physiological processes. In particular, Magnetic

S. Cetin-Karayumak et al. (Eds.): CDMRI 2022, LNCS 13722, pp. 26–37, 2022.
https://doi.org/10.1007/978-3-031-21206-2_3

Resonance Imaging (MRI) is a formidable imaging technique providing a plethora of contrasts through different modalities which can be used to quantify specific features of biological tissues non-invasively. As such, the MRI use in clinics has transformed the diagnosis, management and treatment of disease.

Recent developments in both MRI scanners' hardware [6] and methods (e.g., [14,26]) have pushed further the capabilities of medical imaging but also opened new challenges in terms of storage and sharing requirements of ever larger MRI datasets. For example, advanced 4D diffusion-weighted MRI (dMRI) datasets can require $\sim$ 100 MB - 10 GB, depending on the spatial resolution and the number of measurements. Moreover, large imaging studies scale this figure up to $>$ 10 TB. An example is the UK-Biobank initiative (https://www.ukbiobank.ac.uk) which aims to collect extensive multi-modal MRI datasets from about 500,000 participants. For each participant, the size of a dMRI dataset is 550 MB, leading to a required storage of 275 TB for the dMRI data alone. Clearly, the high-dimensional nature of these data poses a challenge in terms of storage and sharing capabilities and associated costs and technology needs, requiring appropriate algorithms able to represent the information in a low-dimensional space.

In this respect, neural networks have been recently shown to be ideal tools to map pixel/voxel locations to image features. The learnt mappings are typically called implicit neural representations and have been used to represent images [18], 3D scenes [17], videos [10] audio [16], and more. In particular, neural networks with periodic activation functions have been recently proposed as powerful tools for implicit neural representation of images [16] and can be used for efficient lossy compression of 2D images [4]. With this strategy, there is no need to generalize to out-of-distribution samples: the compression procedure coincides with the training itself, with reasonable time/energy consumption on ordinary workstation or even low-power devices. A drawback of this approach is that the resulting compression is lossy, meaning that the compression comes at the cost of losing a fraction of the information. The suitability of lossy compression in clinical applications has been widely investigated for CT, but relatively less for MRI. The few studies concerning MRI have concluded that JPEG and JPEG2000 lossy compression ratio up to 25 preserves diagnostic accuracy and perceived image quality [19]. However, some of the issues arising from the compression of CT data may be relevant to MR imaging. A wide range of studies have investigated mostly JPEG2000 compression of CT data and have concluded that acceptable lossy compression ratios range from 4 to 20 [1,7,9,13,25].

The aim of this work is to evaluate how neural networks with periodic activation functions can be effectively leveraged to compress large multidimensional medical image datasets, with proof-of-concept application on 4D dMRI. By extending the approach proposed in [4,16] to 4D images, we quantitatively investigate the impact of lossy compression and show how any given 4D dMRI dataset can be accurately represented through the parameters of a sinusoidal activation network, achieving a data lossy compression rate about 10 times higher than the current standard lossless DEFLATE algorithm [3]. We envision that this architecture will aid not only compression but also security and anonymization and could be applied to other image modalities, beyond this demonstration on dMRI.

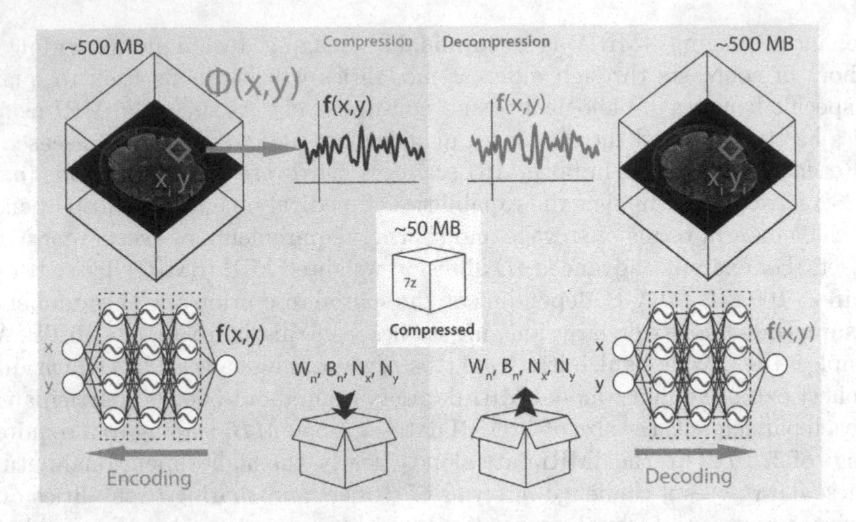

Fig. 1. Compression and decompression procedure. Overview of the main steps involved in the proposed compression and decompression procedure.

2 Methodology

The proposed compression approach uses the SIREN (sinusoidal representation networks) architecture [16], which consists of a multilayer perceptron (MLP) with sine activation functions for implicit neural representations. The overall compression and decompression procedure is outlined in Fig. 1. As proposed in [4], the encoding step leverages the overfitting of the MLP to a given image, quantizing its weights and biases and storing those as a compressed reconstruction of the image. As a further lossless compression step, the network parameters are archived with a Lempel-Ziv-Markov chain algorithm (https://7-zip.org). At decoding time, the MLP is initialized with the stored parameters and evaluated using pixel locations as input to reconstruct the image.

We generalized this formulation for 4D datasets, where the goal becomes learning the mapping between the 3D coordinates of a given voxel and its modality-specific signal variations. As benchmark, we compared the performances of this approach with those from standard MLPs with ReLU and Tanh activation functions, widely used for implicit neural representation tasks [16]. To quantify their performances, we relied on several measures, including mean squared error (MSE), peak signal-to-noise ratio (PSNR) and structure similarity index (SSIM). Furthermore, we compared the loss of information from the proposed SIREN approach with benchmark lossy compression JPEG2000 in terms of relative error of estimated indices from tensor and spherical harmonics representations. Finally, as reference for the acceptable margins of these relative errors, we compared them with those resulting from two commonly used pre-processing steps in dMRI: MP-PCA denoising [23] and smoothing with a Gaussian kernel (FWHM = 1.5).

SIREN Mathematical Formulation. We propose to approach the compression problem of a multi-dimensional image as finding an implicit representation $\mathbf{\Phi}$ between its coordinates $\mathbf{x}_i$ and its features $\mathbf{y}_i = \mathbf{f}(\mathbf{x}_i)$. In the application to dMRI images mentioned in the introduction, each voxel is associated with a distribution of N_{meas} directional values. To approximate $\mathbf{\Phi}$, we seek to solve the optimization problem with loss function $\mathcal{L} = \sum \|\mathbf{\Phi}(\mathbf{x}_i) - \mathbf{f}(\mathbf{x}_i)\|^2$. Leveraging previous work on implicit representation, we define $\mathbf{\Phi}$ through a neural network with periodic activation function [16]:

$$\mathbf{\Phi}(\mathbf{x}_i) = \mathbf{W}_n(\phi_{n-1}\circ\phi_{n-2}\circ\phi_0)((x))+\mathbf{b}_n, \qquad \mathbf{x}_i \mapsto \phi_i(\mathbf{x_i}) = sin(\mathbf{W}_i\mathbf{x}_i+\mathbf{b}_i) \quad (1)$$

where ϕ_i represents the ith layer of the network, $\mathbf{W}_i$ is the weight matrix, and $\mathbf{b}_i$ is the bias term. The main advantage of periodic activation functions is that the derivatives remain well behaved for any weight configuration, and as a result it is possible to learn not only the mapping $\mathbf{\Phi}$ but also all its derivatives. One caveat of these activation functions is that due to their periodic nature, catastrophic forgetting phenomena during training can occur. This issue can be easily overcome initializing the weights from a uniform distribution between $-\sqrt{(6/n)}$ and $\sqrt{(6/n)}$, where n is the number of inputs to each activation unit: this constrain ensures that the periodic activation input has a normal distribution with a unitary standard deviation [16]. In principle, this formulation can be applied to a tridimensional volume where $\mathbf{x}_i = (x_i, y_i, z_i)$. However, in this case the number of parameters needed to properly learn the mapping $\mathbf{\Phi}$ could become dramatically higher and poses a burden on the subsequent implementation. For this reason, a more feasible approach could be to learn $\mathbf{\Phi}$ for each slice relying on its bidimensional coordinates $\mathbf{x}_i = (x_i, y_i)$. Here, we implemented and compared both the approaches. A further observation is that combining SIREN and ReLU approaches could lead to the best of both worlds, hence we also experimented with hybrid architectures.

Implementation. The networks were implemented using PyTorch, extending previous work made available from Sitzmann and colleagues (https://github.com/vsitzmann/siren). All the code is publicly available at the following GitHub repository: https://github.com/palombom/SirenMRI. For our experiments, we used five subjects from the publicly available MGH HCP Adult Diffusion dataset (https://db.humanconnectome.org/). The model fitting was performed using a NVIDIA Titan XP GPU for 2D architectures, while for 3D ones up to four NVIDIA Tesla V100 SXM2 GPUs were used in parallel. The training is performed in a self-supervised way, so the training set consists of all the voxels within a dMRI dataset (in this case $140 \times 140 \times 96 = 1881600$). There is no split in training/validation/testing sets (and no need for it) as our goal is to overfit the input data. A new network is trained for each dataset. During training, which corresponds to the encoding or compression phase, we update the network's weights and biases using back-propagation and mean squared error as loss function (calculated between the input data and the network's predictions) which is minimized using the ADAM algorithm (2000 epochs, determined experimentally as trade-off between highest peak SNR, highest compression ratio and fastest training). The learning rates were determined experimentally as $3 \cdot 10^{-4}$ and $2 \cdot 10^{-4}$ for 2D and 3D architectures, respectively. The training (i.e. compression) for the whole dMRI dataset took on average 13 min, while the prediction (i.e. decompression) took about 2 s. We explored the impact of both deeper and wider architectures testing networks with 3/4/5 layers and 128/256/512 units per layer (1024 units for 3D architectures).

Assessing Networks' Performances. In the first experiment, we assessed the performance of each network. We quantified the average MSE, PSNR and compression ratio for a representative dMRI dataset (Subject ID: MGH1001) as:

$$MSE = \frac{1}{N_{voxels}} \frac{1}{N_{meas}} \sum_{i=1}^{N_{voxels}} \sum_{j=1}^{N_{meas}} [\hat{S}_j^{ground-truth}(\mathbf{x}_i) - \hat{S}_j^{decomp}(\mathbf{x}_i)]^2 \quad (2)$$

$$PSNR = 20\ log_{10}(\frac{1}{\sqrt{(MSE)}}) \quad (3)$$

where $\hat{\mathbf{S}}(\mathbf{x}_i)$ is the vector of N_{meas} dMRI signals from the voxel at location $\mathbf{x}_i$, normalized between 0 and 1. For each decompressed image, we also calculated voxel-wise the relative error and the local SSIM with respect to the ground-truth using the windowing approach proposed in [24] and implemented in scikit-image (https://scikit-image.org). We finally computed the compression ratio as the ratio between the uncompressed and compressed image file sizes.

Evaluating Compression Quality and Accuracy. In the second experiment, we quantitatively assessed the quality of the compression obtained by the network comprised of 3 layers and 256 units which showed a good compromise between PSNR (> 36) and compression ratio ($\sim$10). Specifically, we quantified the accuracy of the compression for metrics of interest in dMRI applications, such as the diffusion tensor [8], the spherical harmonics coefficients [2] and the fibre orientation distribution function (fODF) [21]. From each of the five subjects in our dataset, the diffusion tensor and related rotational invariant metrics fractional anisotropy (FA) and mean diffusivity (MD) were estimated using only the shell at $b = 1,000$ s/mm^2 with MRtrix3 (https://www.mrtrix.org) [22]. The spherical harmonics coefficients up to the 4-th order were estimated using only the shell at $b = 5,000$ s/mm^2 with MRtrix3. The corresponding rotational invariant metrics RISH0 and RISH2 were then computed according to [12]. Finally, the fODF was estimated using the constrained spherical deconvolution algorithm as implemented in MRtrix3 using only the shell at $b = 5,000$ s/mm^2.

3 Results

Networks' Performances. In Fig. 2a we show the performances during training for the representative networks comprised of 3 layers and 256 units, demonstrating that all the networks converged to their best PSNR/MSE. We note that introducing a ReLU activation function after the last layer of a SIREN-2D network (SIREN-2D-relu) stabilizes the training and slightly improves the

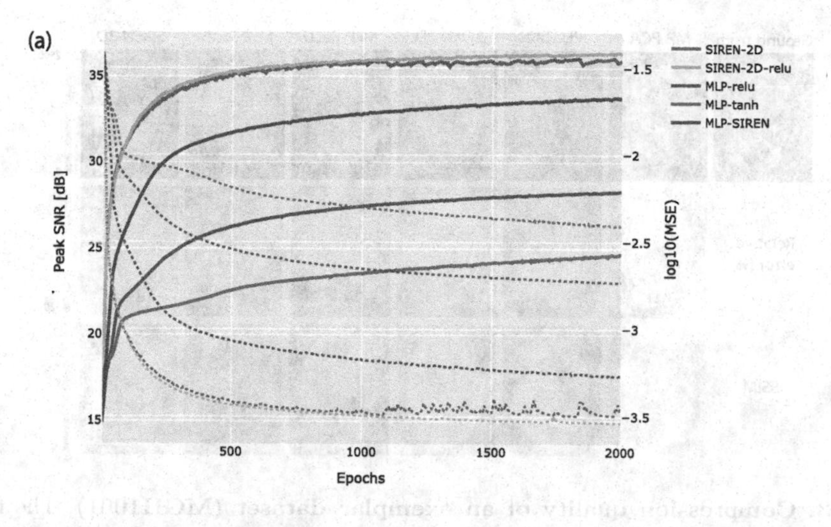

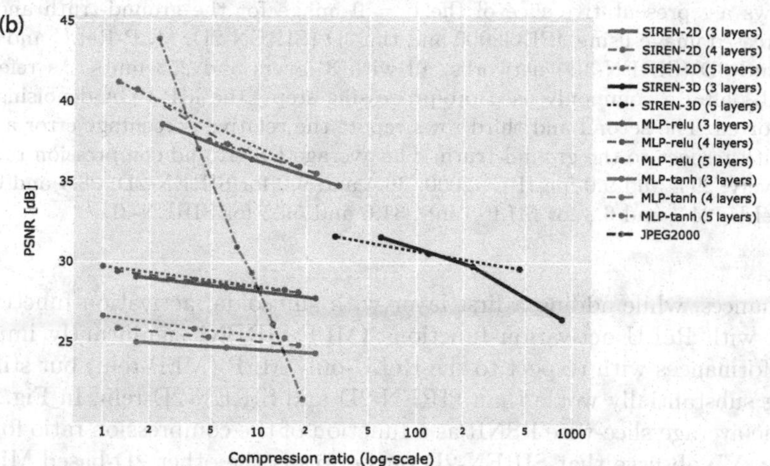

Fig. 2. Networks' performances. *(a)* PSNR (solid lines) and log10 (MSE) (dashed lines) as a function of training epochs for the different networks and the exemplar configuration of 3 layers and 256 units per layer, chosen as overall good compromise between the achievable PSNR and corresponding compression ratio. *(b)* PSNR as a function of compression ratio for the different networks used and JPEG2000. In each curve, the points represent an increasing number of units per layer when going from higher to lower compression ratios. Specifically, for the 2D networks (SIREN-2D, MLP-ReLU and MLP-Tanh) these correspond to: 128/256/512 units, while for the 3D one (SIREN-3D): 256/512/1024 units.

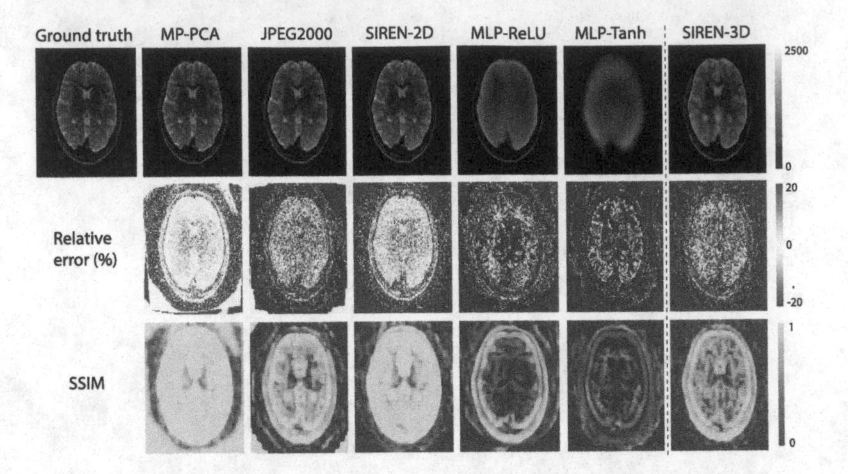

Fig. 3. Compression quality of an exemplar dataset (MGH1001). The first row shows a representative slice of the b = 0 image for the ground-truth and each decompressed image using JPEG2000 and the 2D (SIREN-2D, MLP-ReLU and MLP-Tanh) and 3D (SIREN-3D) networks, all with 3 layers and 256 units. As reference, the results after a commonly used pre-processing step - the MP-PCA denoising - are also reported. The second and third rows report the relative percentage error and the SSIM with respect to the ground-truth. The average PSNR and compression ratio are respectively: 29.3 and 9.0 for JPEG2000; 36.4 and 9.0 for SIREN-2D; 28.4 and 6.2 for MLP-ReLU; 24.7 and 6.3 for MLP-Tanh; 31.6 and 59.5 for SIREN-3D.

performances, while adding a first layer with sinusoidal activation functions to a MLP with ReLU activation functions (MLP-SIREN) significantly improves the performances with respect to the ReLU-only MLP (MLP-relu) but still performing substantially worse than SIREN-2D and SIREN-2D-relu. In Fig. 2b we show the average slice-wise PSNR as a function of the compression ratio for each network. We observe that SIREN-2D outperforms the other 2D-based MLPs in terms of both PSNR and compression ratio; and the JPEG2000 at compression ratios higher than 3. With respect to the 3D implementation SIREN-3D, SIREN-2D still provides higher PSNR, although at a reduced compression ratio.

Compression Quality and Accuracy. In Fig. 3 we compare the compression quality for the commonly used lossy compression JPEG2000 and all the networks comprised of 3 layers and 256 units using the b = 0 image from the subject MGH1001. We find that SIREN-2D provides the lowest relative error and the highest SSIM among the lossy compressions. It is worthwhile noting that masking the brain, if the background is not of interest, can lead to further compression of a factor 2 to 3. In terms of diffusion metrics, in Table 1 we report the comparison of the relative error obtained by SIREN-2D with 3 layers and 256 units with the relative error obtained by JPEG2000 and two pre-processing steps commonly used in dMRI data analysis: MP-PCA denoising and Gaussian smoothing. We found that the relative error from SIREN-2D is close to the relative error from MP-PCA denoising, and from about 5 to 20 times lower

Table 1. Relative error of diffusion-based metrics (MD: mean diffusivity and FA: fractional anisotropy from the data at b = 1000 s/mm^2; RISH: rotiationally invariant spherical harmonics from the data at b = 5000 s/mm^2) for one representative subject after the two exemplar pre-processing steps: MP-PCA denoising and smoothing with a Gaussian kernel; and compression/decompression using two lossy compression methods: the SIREN-2D network with 3 layers and 256 units, and the JPEG2000 algorithm with equivalent compression ratio. The mean (± std) values were calculated using the voxels in the white (WM) and gray (GM) matter, and cerebrospinal fluid (CSF) masks.

Method		**MD**		
		WM	*GM*	*CSF*
Pre-Processing:	*MP-PCA*	1.98%(±12.54%)	0.75%(±1.27%)	1.64%(±2.55%)
	Smoothing	3.02%(±44.87%)	2.89%(±13.03%)	−2.29%(±3.73%)
Compression:	*JPEG2000*	5.71%(±132.38%)	3.89%(±29.59%)	−4.01%(±10.22%)
	SIREN-2D	2.16%(±83.81%)	1.02%(±7.45%)	1.56%(±3.31%)
Method		**FA**		
		WM	*GM*	*CSF*
Pre-Processing:	*MP-PCA*	0.44%(±4.58%)	−0.12%(±9.10%)	−1.10%(±18.05%)
	Smoothing	−5.36%(±5.75%)	−8.68%(±8.34%)	−8.49%(±11.18%)
Compression:	*JPEG2000*	−7.98%(±21.92%)	−0.09%(±36.63%)	5.39%(±40.73%)
	SIREN-2D	−0.72%(±9.82%)	−1.03%(±16.13%)	−0.07%(±22.81%)
Method		**RISH0**		
		WM	*GM*	*CSF*
Pre-Processing:	*MP-PCA*	0.01%(±0.80%)	0.01%(±0.72%)	0.01%(±0.66%)
	Smoothing	−1.83%(±2.63%)	0.78%(±5.24%)	6.16%(±7.99%)
Compression:	*JPEG2000*	−1.54%(±5.24%)	0.69%(±11.89%)	11.02%(±21.09%)
	SIREN-2D	−0.50%(±3.39%)	0.24%(±5.31%)	0.99%(±7.44%)
Method		**RISH2**		
		WM	*GM*	*CSF*
Pre-Processing:	*MP-PCA*	−0.73%(±10.91%)	−6.94%(±41.90%)	−10.95%(±45.00%)
	Smoothing	−11.41%(±8.18%)	−19.78%(±16.17%)	−27.42%(±21.47%)
Compression:	*JPEG2000*	−8.60%(±18.78%)	−6.32%(±62.54%)	10.25%(±106.32%)
	SIREN-2D	−4.92%(±17.55%)	−6.80%(±53.25%)	2.52%(±92.75%)

Table 2. Relative error for five subject after compression/decompression of diffusion-based metrics (MD and FA from the data at b = 1000 s/mm^2; RISH from the data at b = 5000 s/mm^2) for the SIREN-2D network with 3 layers and 256 units. The mean (± std) values were calculated using all the voxels in the brain mask, including CSF.

Subject ID	MD	FA	RISH0	RISH2
MGH1001	−0.69%(±5.37%)	−3.93%(±11.96%)	−0.06%(±4.54%)	−13.99%(±28.00%)
MGH1002	−0.72%(±5.57%)	−3.89%(±12.17%)	−0.11%(±4.66%)	−14.17%(±28.14%)
MGH1003	−0.87%(±4.99%)	−4.37%(±12.06%)	−0.13%(±4.72%)	−15.56%(±29.64%)
MGH1004	−0.82%(±5.48%)	−3.94%(±12.32%)	−0.09%(±4.60%)	−14.51%(±29.04%)
MGH1005	−0.65%(±5.23%)	−3.65%(±11.56%)	−0.10%(±4.49%)	−11.33%(±25.60%)

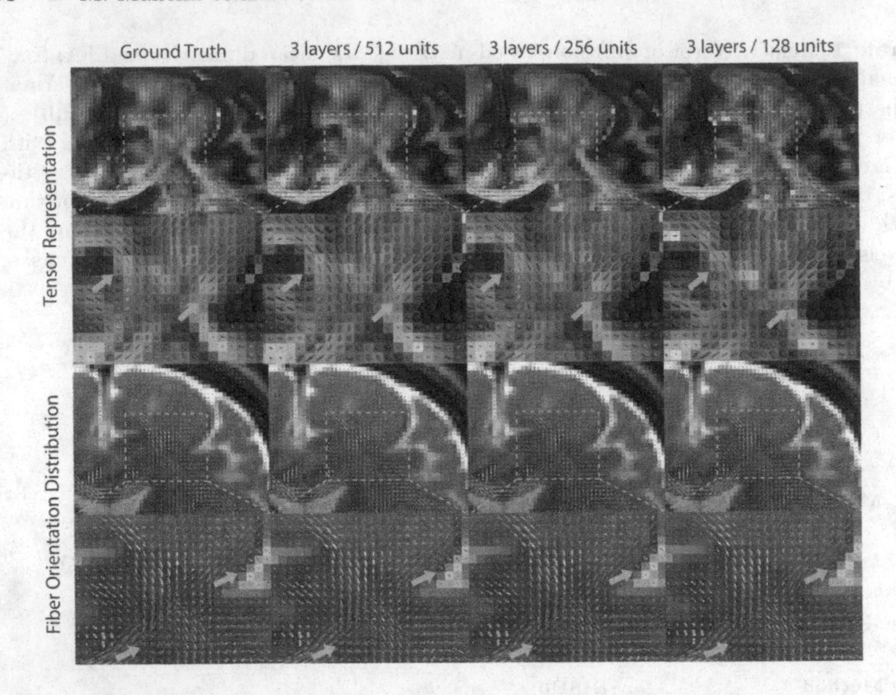

Fig. 4. Reconstructed diffusion tensor and fODF for an exemplar dataset (MGH1001) and different SIREN-2D architectures. Arrows point at regions where differences are more evident, especially for the 3 layers/128 units case.

than smoothing and JPEG2000 compression. In Table 2 we report the relative error after compression and subsequent decompression for FA, MD, RISH0 and RISH2, obtained with the same SIREN-2D network with 3 layers and 256 units across five subjects. Consistently, we find that the relative error on FA is $< 5\%$, on MD and RISH0 $< 1\%$ and on RISH2 $\sim 15\%$. Finally, in Fig. 4 we show the impact of the number of units for the SIREN-2D network with 3 layers on the diffusion tensor and the fODF. As expected, SIREN-2D with 512 units provides the most accurate result, however its compression ratio is only ~ 2. Very small differences can be seen between this network and SIREN-2D with 256 units, while some relevant differences are noticeable with SIREN-2D with 128 units.

Summary. Overall, our results show that SIREN-2D with 3 layers/256 units is a good compromise to achieve a compression ratio of ~ 10 with high accuracy. Importantly, the relative error obtained with this approach is a) smaller than commonly used lossy JPEG2000 compression; b) much smaller than commonly used pre-processing steps of Gaussian smoothing and c) similar to the commonly used pre-processing MP-PCA denoising, suggesting minimal loss of information within the currently accepted levels for clinical applications and utility.

4 Discussion and Conclusion

The results of our experiments show how periodical activation functions can dramatically change the performances of MLPs, providing a powerful tool for compression purposes. This is thanks to the ability of periodical activation functions to learn the non-linear mapping between the position of each voxel in a 4D dMRI image and the corresponding features of the dMRI signal *as well as* all its derivatives [16]. Here, we have demonstrated two important aspects for the design and training of sinusoidal activation networks: a) a single layer with sinusoidal activation functions in a conventional MLP with ReLU (or Tanh) activation functions improves its performances but is not enough to fully exploit SIREN's properties; b) a ReLU activation function after the last layer of a sinusoidal activation network provides more stable convergence to the optimum.

This work is inspired by recent studies showing how SIREN networks can be used to compress several kind of data, including images from different modalities, videos and audios [4,5,16]. However, none of the previous studies have assessed quantitatively the downstream impact of the SIREN lossy compression on medical images analysis, and to what extent this is acceptable. To the best of our knowledge, this is the first work to quantitatively evaluate the performance of SIREN for compression of multidimensional medical imaging modalities. Our quantitative results highlight how the compression procedure, despite lossy, preserves the measures underlying the images with high accuracy. In fact, the estimated relative error subsequent to SIREN compression (up to compression ratio 10) were very similar to the error from commonly used pre-processing steps, such as MP-PCA denoising. This suggests minimal loss of information within the currently accepted levels for clinical applications and utility.

It is worthwhile noting that, for compression ratios ≤ 2, JPEG2000 outperforms SIREN-2D(-relu) in terms of peak SNR and it would be a better choice for the lossy compression. For higher compression ratios, SIREN-2D(-relu) should be used instead, as it largely outperforms JPEG2000. Also, on average, the lossy compression algorithms tested here showed large standard deviations which may require further investigations. With respect to JPEG2000, SIREN-2D showed consistently lower standard deviations, but larger than MP-PCA denoising.

In addition to the good compromise between PSNR and compression ratio, the proposed implementation does not require to generalize as the goal of this approach is to overfit the data itself. Therefore it does not require large datasets for training and testing, as in other deep learning based compression methods (e.g. autoencoders [20]). Our results from different subjects confirm the stability of the compression performances. Moreover, given the relatively shallow and simple architecture of the SIREN network, the proposed method can be implemented on ordinary workstations, offering a valid low-energy/low-cost alternative to more demanding deep learning solutions. Future iterations could take advantage of multiple low-energy devices to parallelize the compression, reducing even more the associated costs. In fact, the main cost behind this approach is related to the time consumption of the compression phase, e.g. 8 s per slice with the 3-layers/256-units SIREN-2D on a NVIDIA Titan XP GPU (our datasets are comprised of 96 slices). Nonetheless, in the application scenario we envision, the compression needs to be done only once and could potentially be done overnight, even directly on the MRI scanner workstation.

It is finally worth noting that, as most multi-dimensional medical imaging modalities are inherently tridimensional, a 3D architecture would be a tempting choice. However, as observed in our experiments, the number of parameters necessary to obtain reasonable PSNR would be prohibitive: even using multiple GPUs in parallel, it was not possible to outperform the simpler 2D implementation. We refer the reader to recent works explaining further the challenges involving 3D compression and possible solutions [5,11]. Potential future directions could explore patch-based approaches, as indeed recently explored in these works. Future work will also explore the performance of the proposed compression methods on multidimensional MRI data entailing the acquisition of diffusion images with different TE, TR, etc., such as the MUDI dataset [15].

Acknowledgements. MM is supported by the Wellcome Trust through a Sir Henry Wellcome Postdoctoral Fellowship (213722/Z/18/Z). MP is supported by the UKRI Future Leaders Fellowship MR/T020296/2.

References

1. Cosman, P.C., et al.: Thoracic ct images: effect of lossy compression on diagnostic accuracy. Radiology 190 (1994)
2. Descoteaux, M., Angelino, E., Fitzgibbons, S., Deriche, R.: Regularized, fast, and robust analytical q-ball imaging. Magnetic Resonance Med. Official J. Int. Soc. Magnetic Resonance Med. **58**(3), 497–510 (2007)
3. Deutsch, P.: Rfc 1951: Deflate compressed data format specification version 1.3 (1996)
4. Dupont, E., Goliński, A., Alizadeh, M., Teh, Y.W., Doucet, A.: Coin: compression with implicit neural representations. arXiv preprint arXiv:2103.03123 (2021)
5. Dupont, E., Loya, H., Alizadeh, M., Goliński, A., Teh, Y.W., Doucet, A.: Coin++: Data agnostic neural compression. arXiv preprint arXiv:2201.12904 (2022)
6. Jones, D.K., et al.: Microstructural imaging of the human brain with a 'super-scanner': 10 key advantages of ultra-strong gradients for diffusion mri. Neuroimage **182**, 8–38 (2018)
7. Ko, J.P., et al.: Wavelet compression of low-dose chest ct data: effect on lung nodule detection. Radiology **228**(1), 70–75 (2003)
8. Le Bihan, D., et al.: Diffusion tensor imaging: concepts and applications. J. Magnetic Resonance Imaging Official J. Int. Soc. Magnetic Resonance Med. **13**(4), 534–546 (2001)
9. Lee, K.H., et al.: Irreversible jpeg 2000 compression of abdominal ct for primary interpretation: assessment of visually lossless threshold. Eur. Radiol. **17**(6), 1529–1534 (2007)
10. Li, Z., Niklaus, S., Snavely, N., Wang, O.: Neural scene flow fields for space-time view synthesis of dynamic scenes. In: Proceedings of the IEEE/CVF Conference on Computer Vision and Pattern Recognition, pp. 6498–6508 (2021)

11. Mehta, I., Gharbi, M., Barnes, C., Shechtman, E., Ramamoorthi, R., Chandraker, M.: Modulated periodic activations for generalizable local functional representations. In: Proceedings of the IEEE/CVF International Conference on Computer Vision, pp. 14214–14223 (2021)
12. Mirzaalian, H., et al.: Inter-site and inter-scanner diffusion mri data harmonization. Neuroimage **135**, 311–323 (2016)
13. Ohgiya, Y., et al.: Acute cerebral infarction: effect of jpeg compression on detection at ct. Radiology **227**(1), 124–127 (2003)
14. Palombo, M., et al.: Sandi: a compartment-based model for non-invasive apparent soma and neurite imaging by diffusion mri. Neuroimage **215**, 116835 (2020)
15. Pizzolato, M., et al.: Acquiring and predicting multidimensional diffusion (MUDI) data: an open challenge. In: Bonet-Carne, E., Hutter, J., Palombo, M., Pizzolato, M., Sepehrband, F., Zhang, F. (eds.) Computational Diffusion MRI. MV, pp. 195–208. Springer, Cham (2020). https://doi.org/10.1007/978-3-030-52893-5_17
16. Sitzmann, V., Martel, J., Bergman, A., Lindell, D., Wetzstein, G.: Implicit neural representations with periodic activation functions. Adv. Neural.Inf. Process. Syst. **33**, 7462–7473 (2020)
17. Sitzmann, V., Zollhöfer, M., Wetzstein, G.: Scene representation networks: Continuous 3d-structure-aware neural scene representations. In: Advances in Neural Information Processing Systems 32 (2019)
18. Stanley, K.O.: Compositional pattern producing networks: a novel abstraction of development. Genet. Program Evolvable Mach. **8**(2), 131–162 (2007)
19. Terae, S., et al.: Wavelet compression on detection of brain lesions with magnetic resonance imaging. J. Digit. Imaging **13**(4), 178–190 (2000)
20. Theis, L., Shi, W., Cunningham, A., Huszár, F.: Lossy image compression with compressive autoencoders. arXiv preprint arXiv:1703.00395 (2017)
21. Tournier, J.D., Calamante, F., Connelly, A.: Robust determination of the fibre orientation distribution in diffusion mri: non-negativity constrained super-resolved spherical deconvolution. Neuroimage **35**(4), 1459–1472 (2007)
22. Tournier, J.D., et al.: Mrtrix3: a fast, flexible and open software framework for medical image processing and visualisation. Neuroimage **202**, 116137 (2019)
23. Veraart, J., Novikov, D.S., Christiaens, D., Ades-Aron, B., Sijbers, J., Fieremans, E.: Denoising of diffusion mri using random matrix theory. Neuroimage **142**, 394–406 (2016)
24. Wang, Z., Bovik, A.C., Sheikh, H.R., Simoncelli, E.P.: Image quality assessment: from error visibility to structural similarity. IEEE Trans. Image Process. **13**(4), 600–612 (2004)
25. Yamamoto, S., et al.: Evaluation of compressed lung ct image quality using quantitative analysis. Radiat. Med. **19**(6), 321–342 (2001)
26. Zhang, H., Schneider, T., Wheeler-Kingshott, C.A., Alexander, D.C.: Noddi: practical in vivo neurite orientation dispersion and density imaging of the human brain. Neuroimage **61**(4), 1000–1016 (2012)

Correction of Susceptibility Distortion in EPI: A Semi-supervised Approach with Deep Learning

Antoine Legouhy[1,2](✉), Mark Graham[3], Michele Guerreri[1,2], Whitney Stee[4,5], Thomas Villemonteix[4,5], Philippe Peigneux[4,5], and Hui Zhang[1]

[1] Centre for Medical Image Computing, Department of Computer Science, University College London, London, UK
a.legouhy@ucl.ac.uk

[2] AINOSTICS Ltd., Manchester, UK

[3] Department of Biomedical Engineering, School of Biomedical Engineering and Imaging Sciences, King's College London, London, UK

[4] UR2NF-Neuropsychology and Functional Neuroimaging Research Unit Affiliated at CRCN - Centre for Research in Cognition and Neuroscience, Université Libre de Bruxelles (ULB), Brussels, Belgium

[5] UNI-ULB Neuroscience Institute, Université Libre de Bruxelles (ULB), Brussels, Belgium

Abstract. Echo planar imaging (EPI) is the most common approach for acquiring diffusion and functional MRI data due to its high temporal resolution. However, this comes at the cost of higher sensitivity to susceptibility-induced B_0 field inhomogeneities around air/tissue interfaces. This leads to severe geometric distortions along the phase encoding direction (PED). To correct this distortion, the standard approach involves an analogous acquisition using an opposite PED leading to images with inverted distortions and then non-linear image registration, with a transformation model constrained along the PED, to estimate the voxel-wise shift that undistorts the image pair and generates a distortion-free image. With conventional image registration approaches, this type of processing is computationally intensive. Recent advances in unsupervised deep learning-based approaches to image registration have been proposed to drastically reduce the computational cost of this task. However, they rely on maximizing an intensity-based similarity measure, known to be suboptimal surrogate measures of image alignment. To address this limitation, we propose a semi-supervised deep learning algorithm that directly leverages ground truth spatial transformations during training. Simulated and real data experiments demonstrate improvement to distortion field recovery compared to the unsupervised approach, improvement image similarity compared to supervised approach and precision similar to TOPUP but with much faster processing.

Keywords: Deep learning registration · Susceptibility distortion correction · Semi-supervised learning

S. Cetin-Karayumak et al. (Eds.): CDMRI 2022, LNCS 13722, pp. 38–49, 2022.
https://doi.org/10.1007/978-3-031-21206-2_4

1 Introduction

Echo planar imaging (EPI) is the most common approach for acquiring diffusion and functional MRI data due to its high temporal resolution which both reduces the influence of motion and allows the acquisition of a large number of volumes in a time frame amenable to neuroscientific and clinical research. This is, however, at the cost of higher sensitivity to susceptibility-induced B_0 field inhomogeneities around interfaces of air, bone, and soft tissue. This leads to severe geometric distortions in the form of local expansions or contractions along the phase encoding direction (PED), breaking alignment with the corresponding anatomical scans and corrupting subsequent diffusion model fitting or tractography. Moreover, the effect of EPI susceptibility distortions radiate over the whole image, introducing systematic alterations even far from the apparent hot spots [6].

To tackle this problem, the strategy that has proved most effective is to acquire an extra scan with identical settings, except for an opposite PED [1] (also referred as blip up, blip down). It produces an analogous image with reverse distortion: expansions where there were contractions and vice versa. One can then apply non-linear image registration, with a transformation model constrained along the PED, to estimate the shift in voxel coordinates that undistorts the image pair and generates a distortion-free image. Standard implementations of this strategy, e.g. TOPUP [4] in FSL, are computationally intensive. They rely on traditional image registration methods that align each new pair of images with a separate iterative optimization. A comparison of such algorithms is available in [2]. It was proposed in [3] to synthesize a b = 0 image through deep learning thus allowing to use TOPUP with the economy of the opposite PED acquisition, but with no gain in terms computational time.

Recently, image registration based on deep learning (DL) architectures, convolutional neural networks (CNN) in particular, have been developed. With the investment of an upfront cost during training, test images can be registered in one-shot almost instantaneously. In the same vein as the unsupervised Voxelmorph [16] framework for anatomical images, it has led to the development of fast EPI distortion correction with DL-powered registration. In [8] and [7], similarly to traditional registration, the optimization relies on maximizing an intensity similarity measure, known to be suboptimal surrogate measures of image alignment [9]. Also, [8] does not integrate intensity modulation to account for signal stretchings and pile-ups associated to geometric expansions and contractions. This aspect is given due consideration in [7], but the network is based on a 2D architecture likely to miss volumetric characteristics. In [5], more reliable fiber orientation distribution (FOD) features are used, although this is intended only to be used as secondary correction of the distortion residuals after an external primary one.

To address those limitations, we propose a 3D semi-supervised DL algorithm that directly leverages ground truth spatial transformations during training. We hypothesise that constraining DL model training with the most direct representations of spatial correspondence will significantly improve the fidelity of the recovered spatial correspondence during testing. Also we integrate Jacobian intensity modulation when constructing the undistorted images.

We performed two experiments to evaluate the proposed method against the unsupervised approach similar to [8], but also against a fully transformation supervised one. The first experiment makes use of real data to assess the practical performance. For this experiment, transformations produced by TOPUP are used as ground truth for training and testing. The second experiment makes use of simulated data, generated by DW-POSSUM [1], a realistic Spin-Echo EPI simulator. This sets up a scenario where we have genuine ground truth for both undistorted image and the distortion-inducing deformation, and allows the comparison with TOPUP.

The code used to implement the proposed semi-supervised model, as well as the unsupervised and field supervised ones (and more), is available in the open source *sudistoc*[1] repository.

2 Background

2.1 Distortion Model

Susceptibility-induced EPI geometric distortion is well understood [11]. For such multi-slice acquisitions, distortion due to B_0 field inhomogeneities is negligible along the frequency encoding direction. Its effect can thus be parametrized by a unidirectional deformation field V shifting voxel coordinates x along the PED. For the typical PED from posterior to anterior, we have $T_+(x) = x + V(x)$. If the PED is reversed, an opposite displacement field will result: $T_-(x) = x - V(x)$. Henceforth, we will refer to this as the opposite symmetry constraint for the two displacement fields.

Without loss of generality, we will refer to T_+ as the forward transform, and the corresponding distorted image as I_+. Likewise, T_- will be referred to as the backward transform and the corresponding distorted image as I_-. The latent undistorted image will be denoted as $\hat{I}$. Following [12], it can be expressed both in terms of the forward and backward images following:

$$\begin{cases} \hat{I}(x) \sim J_{T_+} \cdot I_+ \circ T_+(x) \\ \hat{I}(x) \sim J_{T_-} \cdot I_- \circ T_-(x) \end{cases} \tag{1}$$

where J_{T_+} (resp. J_{T_-}) is the Jacobian determinant of T_+ (resp. T_-), $\circ$ denoting composition and $\cdot$ element-wise multiplication. J_T encodes the local expansion (if $|J_T| \in [1, +\infty)$) or contraction (if $|J_T| \in (0, 1]$) properties of the transformation and will modulate the resulting intensities accordingly [13].

2.2 Distortion Correction Using Image Registration

Under the above distortion model, it is evident image registration can be used to estimate the transformations for correcting the distorted image pair. Equation (1) suggests the correction can be formulated as the following image registration problem:

$$\underset{V}{\arg\min}\ \mathrm{C}\left(J_{T_+} \cdot I_+ \circ T_+,\ J_{T_-} \cdot I_- \circ T_-\right) \tag{2}$$

[1] *sudistoc*: https://github.com/CIG-UCL/sudistoc.

where C is a dissimilarity criterion between the two corrected images (from I_+ and I_- respectively). The mean squared error (MSE) between intensities is particularly well suited in this case since we are dealing with the same subject, the same modality, the same acquisition parameters (with the exception of PED); we therefore expect almost identical intensities at endpoint. The sought transformations (T_+ and T_-) are parametrized by V which, even though constrained to be unidirectional, can have a large number of degrees of freedom: up to the number of voxels of the image.

As noted in the introduction, this image registration problem is currently typically solved via computationally intensive iterative optimization, with each new image pair solved completely independently. Recent advances in DL-based image registration have recently been leveraged to substantially accelerate this task by replacing iterative optimization with a one-shot computation [14,15]. The idea behind DL-based image registration is to learn a model that can predict from an image pair the transformation that put them into correspondence. During training, model parameters are tuned to output an optimal transformation, for each training sample, that maximises either some similarity between transformed images (unsupervised) or the resemblance to the corresponding ground truth transformation (supervised). While the training may be computationally expensive, once completed, new image pairs can be registered almost instantaneously. The most popular publicly available implementation is VoxelMorph [16,17] which provides an unsupervised framework, making use of a U-Net convolutional neural network (CNN) architecture. This framework was recently exploited for EPI distortion correction [8]. However it has now been limited to optimisation over purely intensity-based losses and does not embed intensity modulation.

3 Method

We implement the proposed semi-supervised approach using the Voxelmorph framework. The framework must be adapted 1) to predict a spatial transform pair with opposite symmetry, 2) to constrain predicted spatial transforms to be unidirectional along the PED, 3) to support the image registration formulation represented by Eq. (2) that includes Jacobian intensity modulation, 4) to enable semi-supervision with ground truth spatial transforms, and 5) to handle weight maps that modulate the contribution of each voxels according to anatomical regions of interest. Points 3, 4 and 5 are not present in the work from [8]. The details of the model are described below.

3.1 Model Architecture

The model is organised into three sequential blocks (Fig. 1). The first block takes a distorted image pair as input and outputs a unidirectional vector field which is used to produce a forward and a backward transformation. Those, together with the associated input images, are fed to a resampling block that reconstruct undistorted images by interpolating from transformed coordinates.

The CNN block is a U-Net as in VoxelMorph, but its input and output are different. Instead of any arbitrary image pair, it must be given a pair of analogously distorted images $\{I_+, I_-\}$ as input. Instead of a vector field, it outputs a scalar field V characterizing constrained displacements along a single direction (PED). As in VoxelMorph, all the trainable parameters of the model are contained in this block.

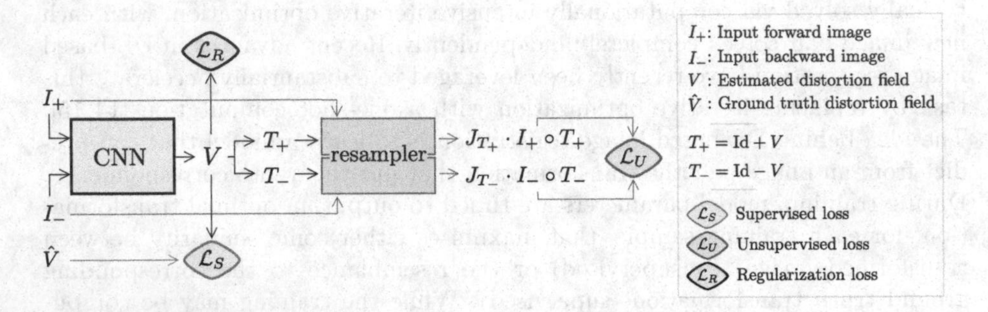

Fig. 1. Diagram of the proposed semi-supervised distortion correction network. Portions specific to unsupervised are highlighted in blue, whereas the ones associated to supervised are in red. (Color figure online)

To preserve the opposite symmetry constraint induced by the PED reversal, the pair of forward and backward transformations $\{T_+, T_-\}$ are built from the estimated field V following: $T_+ = \mathrm{Id} + V$ and $T_- = \mathrm{Id} - V$.

Resampling block implements image warping. Unlike VoxelMorph and [8], our implementation includes intensity modulation, which allows us to account for signal pile-up in the presence of contraction and signal reduction in the presence of expansion. It takes I_+ (resp. I_-) and T_+ (resp. T_-) as input to produce undistorted, intensity-modulated images $J_{T_+} \cdot I_+ \circ T_+$ (resp. $J_{T_-} \cdot I_- \circ T_-$). This requires computing the Jacobian determinant of the transformations.

3.2 Models

The model architecture described above is used to implement an unsupervised and a transformation supervised model, as baselines for comparison, and the proposed semi-supervised model.

A diagram illustrating the different models can be found in Fig. 1. The parts exclusive to the unsupervised model, comparable to [8], are highlighted in blue. It includes the resampling block that unwarps the input images with the estimated field allowing their comparison (unsupervised loss). The red parts are exclusive to the supervised model. It includes the ground truth field that is compared to the estimated one (supervised loss). The black parts are common to both. It includes the distortion field estimation (a regularization loss is computed on it). The semi-supervised model encompasses both the unsupervised and supervised components.

The unsupervised loss $\mathcal{L}_U$ is an image similarity metric between the two undistorted images that have undergone Jacobian intensity modulation. As mentioned in Sect. 2.2 the MSE is well indicated and Eq. 1 leads to:

$$\mathcal{L}_U = \sum_{i=1}^{n} w_i \left(J_{T_+} \cdot I_+ \circ T_+(x_i) - J_{T_-} \cdot I_- \circ T_-(x_i)\right)^2 \tag{3}$$

This loss is only required for the unsupervised and semi-supervised models.

The supervised loss is a distance between the estimated distortion field V and the ground truth one $\hat{V}$. The MSE is also well adapted when dealing with displacement vectors as it corresponds to an average of geometrical distances, it's a direct quantitative measure of the goodness of the registration:

$$\mathcal{L}_S = \sum_{i=1}^{n} w_i \left(V(x_i) - \hat{V}(x_i)\right)^2 \tag{4}$$

This loss is only required for the supervised and the semi-supervised model.

A regularization loss $\mathcal{L}_R$ is also use on the estimated distortion field V to encourage smoothness:

$$\mathcal{L}_R = \frac{1}{n} \sum_{i=1}^{n} \|\nabla_V(x_i)\|_F^2 \tag{5}$$

where ∇_V is the Jacobian of the field V and $\|.\|_F$ is the Frobenius norm. This loss is computed for all models.

For each models, the overall loss is a sum of the ones above, weighted to account for large order-of-magnitude differences in different loss terms.

To prevent the learning process to be influenced by meaningless information from background (which represent the majority of the image!), another kind of weighting, spatially this time, occurs when computing the unsupervised and the supervised losses. It corresponds to the w_i in Eq. 3 and Eq. 4. The contribution of each voxels is modulated such as only areas of interest (typically just the brain) contribute to the loss, ignoring the background. This is not present in Voxelmorph and [8].

4 Evaluations

The idea is to evaluate how well, having processed a portion of a dataset with a regular tool (here TOPUP), one can use those to train a DL model that will rapidly process the rest of the dataset. Corrections from three DL registration approaches are engaged for comparison: unsupervised, transformation supervised and semi-supervised. We will perform experiments on two datasets: 1- A real dataset with TOPUP outputs as ground truths, 2- A synthetic dataset with absolute ground truths allowing to integrate TOPUP to the comparison.

4.1 Datasets

We acquired anatomical and diffusion weighted data as part of an ongoing study investigating memory learning and consolidation. We have, for 60 healthy subjects, T1-weighted images and distorted EPI b = 0 image pairs from opposite PED acquisition that are antero-posterior (AP) and postero-anterior (PA).

Distorted EPI images have been processed through TOPUP to obtain undistorted images and associated distortion fields.

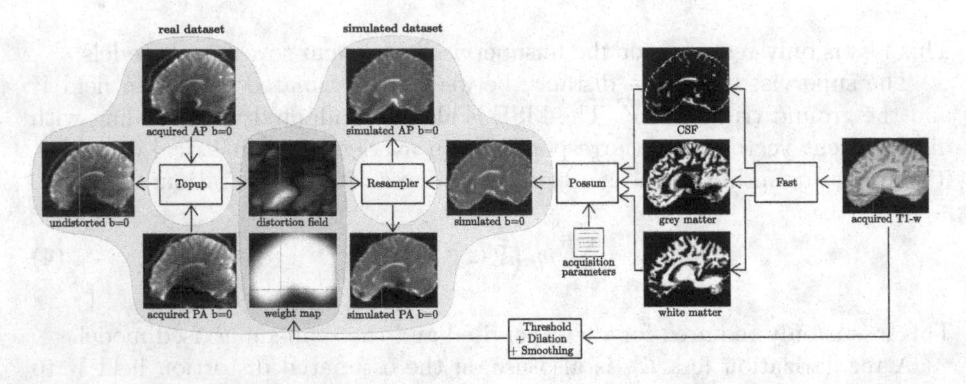

Fig. 2. Preprocessing steps involved in creating the real and the simulated datasets.

In addition, simulated data were produced using DW-POSSUM [1] (an extension of FSL POSSUM [10]), a realistic Spin-Echo EPI simulator. It takes as input tissue segmentations, MR parameters associated with these tissues and a pulse sequence, and produces an EPI image by solving the Bloch equations. We obtained the tissue segmentations by processing the anatomical scans with FSL FAST, producing probability maps for grey matter, white matter and cerebrospinal fluid. We then applied the distortion estimated by TOPUP on the real EPI data of each subject to their corresponding simulated undistorted images to create new synthetic pairs of AP and PA distorted images.

Acquired, TOPUP processed and simulated data have been used to produce two datasets, denoted as real and simulated, in order to cover various experimental configurations. A diagrammatic representation of the processing paths followed to obtain the two datasets is presented in Fig 2.

- In the real dataset: the inputs of the model are the acquired AP and PA images, the ground truth distortion fields used at training and for evaluation are the ones from TOPUP, the ground truth images used only for evaluation are the undistorted images from TOPUP. The advantage of this dataset is that it is made of real data and all the artifacts it implies. The drawback is that TOPUP is used as ground truth and therefore cannot be included in the comparison.

- In the simulated dataset: the inputs of the model are the simulated AP and PA images, the ground truth distortion fields used at training and for evaluation are the ones from TOPUP (same as for the real dataset), the ground truth images used only for evaluation are simulated non-distorted images. The advantage of this dataset is the synthetic, absolute nature of the ground truths allowing any algorithm comparison. The drawback is that the simulation process is not able to reproduce all the artifacts induced by a real acquisition.

For both, the weight maps have been computed by thresholding (binary brain mask), then dilating (3 voxels), then smoothing (Gaussian, σ = 6 mm) the T1-weighted images.

We divided the same way the two datasets into training (n = 40), validation (n = 10) and testing (n = 10) samples. Although it was shown in [16] that Voxelmorph-like architectures can achieve decent registration with only ten or so training subjects.

4.2 Models

We compared 3 deep learning approaches: unsupervised, transformation supervised and semi-supervised. Each model is trained and assessed separately for real and simulated experiments. To form the overall loss, the unsupervised loss was attributed a weight 200 000 (unsupervised and semi-supervised models only), the supervised loss was attributed a weight 300 (supervised and semi-supervised models only) and the regularization loss was attributed a weight 1 (all models). Each model was trained for 1000 epochs, with the epoch giving the best validation loss kept. Learning rate was set to 10^{-4}. The training time for each model was about 3 h on a GPU.

4.3 Assessment Metrics

For both real and simulated data experiments, their respective unseen test data were used to quantitatively assess the performance of the proposed semi-supervised approach against the current unsupervised approach. The assessment made use of the following metrics: 1) Image fidelity: MSE between the estimated undistorted versions of the AP and PA images and the corresponding ground truth undistorted image. 2) Field fidelity: MSE between the estimated distortion field and the corresponding ground truth distortion field (expressed in fraction of the voxel size which is 2 mm). These mean measures were weighted by the weight maps to ignore the background.

5 Results

Table 1 and Fig. 3 summarise the evaluation results for both real and simulated data experiments.

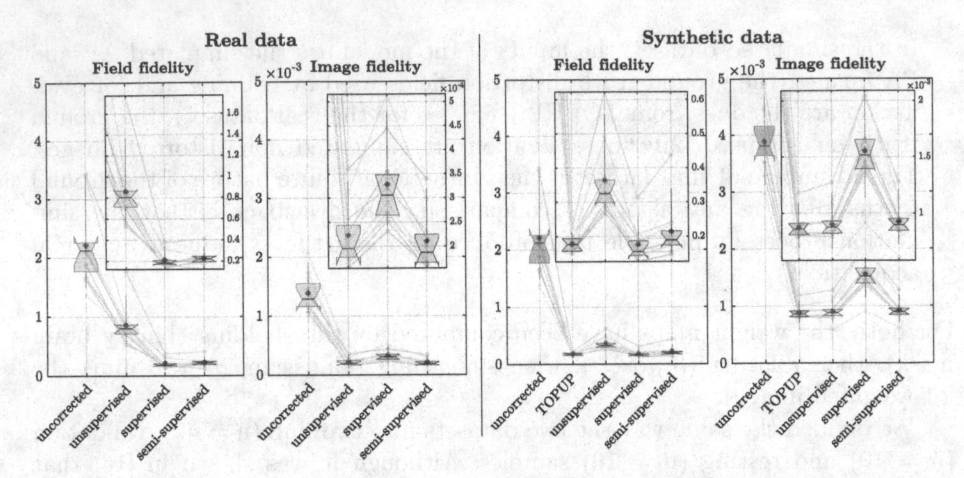

Fig. 3. Fidelity (MSE) to ground truth distortion fields and images for the different models for real data with TOPUP as ground truth and for synthetic data. The field fidelity is expressed as a fraction of the voxel dimension (2 mm). Large frames share the same scale for global overview between experiments whereas yellowish ones are zoomed in for intra-metric, between models comparison. (Color figure online)

- The real dataset experiment uses real acquired EPI images and evaluate how close to TOPUP (ground truth) the different models behave. In terms of field fidelity, supervised and semi-supervised models show a mean voxelique MSE around 0.2 (equivalent to 0.4 mm^2), way better than the unsupervised approach that shows an error about 4 times bigger with also a much higher variance. In terms of image similarity, the unsupervised and the semi-supervised approach show an error equivalent to 2/3 of the one of the supervised approach with akin variance. The semi-supervised approach performs well for both metrics whereas the other models present weaker results in one situation. An example case subject from the testing sample, corrected with the semi-supervised approach, is shown in Fig. 4.
- The synthetic dataset experiment uses simulated EPI images and evaluate how each model and TOPUP are able to retrieve a synthetic ground truth. It quite follow the same trend as above for the deep-learning models. TOPUP, the supervised and the supervised models show similar field fidelity that is better than the one of the unsupervised model. TOPUP, the unsupervised and the supervised models show similar image fidelity that is better than the one of the supervised model. TOPUP and the semi-supervised model performs similarly well for both metrics whereas the supervised and the unsupervised models are weaker in one situation.

Table 1. Fidelity (MSE: mean (std)) to ground truth distortion fields and images for the different models for real data with TOPUP as ground truth and for synthetic data. The field fidelity is expressed as a fraction of the voxel dimension (2 mm).

		Uncorrected	TOPUP	Unsupervised	Supervised	Semi-supervised
Real dataset	Field fidelity	2.21 (0.70)	∅	0.86 (0.34)	0.21 (0.07)	0.22 (0.07)
	Image fidelity	1.39 .10^3 (3.39 .10^4)	∅	2.22 .10^4 (6.76 .10^5)	3.27 .10^4 (9.09 .10^5)	2.10 .10^4 (6.29 .10^5)
Synthetic dataset	Field fidelity	2.21 (0.70)	0.17 (0.02)	0.36 (0.11)	0.17 (0.05)	0.21 (0.08)
	Image fidelity	3.89 .10^3 (5.94 .10^4)	8.73 .10^4 (1.51 .10^4)	8.98 .10^4 (1.82 .10^4)	1.55 .10^3 (3.41 10^4)	9.25 .10^4 (1.81 .10^4)

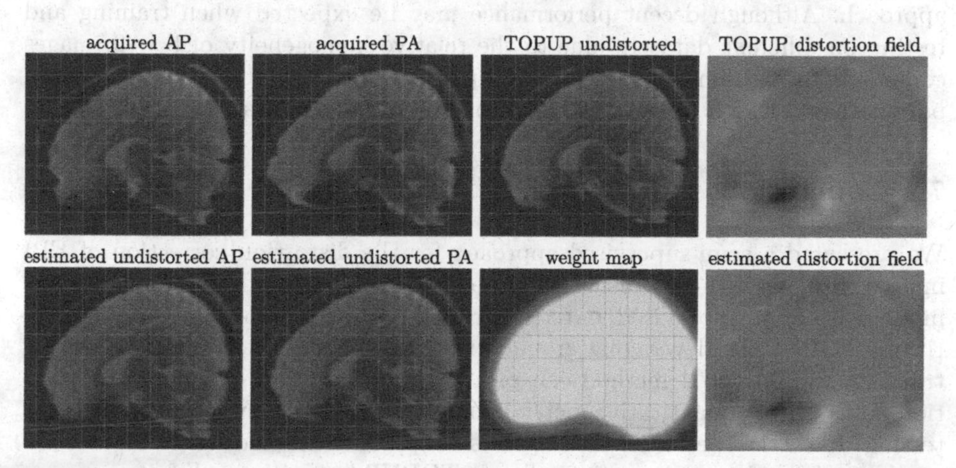

Fig. 4. Example case from the testing sample (unseen) of the real dataset, corrected using the semi-supervised model.

6 Discussion

This study show that, with proper distortion model, one can use deep-learning registration trained on some tens of subjects to rapidly correct a larger set for susceptibility-induced distortion, with results as good as TOPUP. One can use a processed subset of a dataset to include transformation supervision and improve the transformation fidelity which is a direct measure of the quality of the registration. One might have expected the semi-supervised model to perform half-way between its unsupervised and supervised counterparts in terms of field and image fidelity. However, our findings suggest it offers the best of both worlds.

Similar to TOPUP [4], the proposed approach requires reversed PED acquisitions to correct distortion in EPI images trough registration, although this process is performed using a deep-learning architecture, much faster than its classical counterpart. Our approach builds on the purely unsupervised approach in [8], but enhance the training notably using more reliable transformation supervision.

In the broader context of applying deep learning to EPI distortion correction, Schilling et al. [3] proposes a technique that allows TOPUP-based correction in the absence of a second reversed PED acquisition. It works by applying deep learning-based modality transfer to synthesise a distortion-free b = 0 from a T1-weighted scan; the distortion-free b = 0 is then used with the acquired b = 0 for TOPUP processing. Our method can be readily extended to drastically shorten this distortion correction process as well. It could also be used as the primary correction prior to the FOD-based secondary one from [5].

In future work, we plan to evaluate the effect of the training sample size on performance. We also intend to study the generalizability of the proposed approach. Although decent performance may be expected when training and testing on different datasets, due to the relative homogeneity of b = 0 images compared to anatomical ones for example, the high variability of acquisition parameters (TE, TR...) may still lead to sub-optimal results.

7 Conclusion

We presented a semi-supervised approach for the distortion correction of EPI images with opposite PED using DL-based image registration. We compared this model with an unsupervised and a supervised one, as well as a traditional algorithm: TOPUP. By leveraging ground truth distortion transformations during training, the proposed method can produce more accurate estimate of distortion fields (direct quantitative metric) compared to unsupervised approaches at testing. It also outperforms the supervised approach for the image metric. On synthetic data, the results were similar to TOPUP but with much faster computation. The proposed model can typically be trained on a processed subsample of a dataset with an external tool and then be applied to the rest of the dataset to produce distortion correction very efficiently.

References

1. Graham, M.S., Drobnjak, I., Jenkinson, M., Zhang, H.: Quantitative assessment of the susceptibility artefact and its interaction with motion in diffusion MRI. PloS One **12**(10), e0185647 (2017)
2. Gu, X., Eklund, A.: Evaluation of six phase encoding based susceptibility distortion correction methods for diffusion MRI. Front. Neuroinform. **13**, 76 (2019)
3. Schilling, K.G., et al.: Distortion correction of diffusion weighted MRI without reverse phase-encoding scans or field-maps. PLoS One **15**(7), e0236418 (2020)
4. Andersson, J.L., Skare, S., Ashburner, J.: How to correct susceptibility distortions in spin-echo echo-planar images: application to diffusion tensor imaging. Neuroimage **20**(2), 870–888 (2003)
5. Qiao, Y., Shi, Y.: Unsupervised deep learning for FOD-based susceptibility distortion correction in diffusion MRI. IEEE Trans. Med. Imaging **41**(5), 1165–1175 (2022)

6. Begnoche, J.P., Schilling, K.G., Boyd, B.D., Cai, L.Y., Taylor, W.D., Landman, B.A.: EPI susceptibility correction introduces significant differences far from local areas of high distortion. Magn. Reson. Imaging **92**, 1–9 (2022)
7. Zahneisen, B., Baeumler, K., Zaharchuk, G., Fleischmann, D., Zeineh, M.: Deep flow-net for EPI distortion estimation. Neuroimage **217**, 116886 (2020)
8. Duong, S.T.M., Phung, S.L., Bouzerdoum, A., Schira, M.M.: An unsupervised deep learning technique for susceptibility artifact correction in reversed phase-encoding EPI images. Magn. Reson. Imaging **71**, 1–10 (2020)
9. Rohlfing, T.: Image similarity and tissue overlaps as surrogates for image registration accuracy: widely used but unreliable. IEEE Trans. Med. Imaging **31**(2), 153–163 (2012)
10. Drobnjak, I., Gavaghan, D., Süli, E., Pitt-Francis, J., Jenkinson, M.: Development of a functional magnetic resonance imaging simulator for modeling realistic rigid-body motion artifacts. Magn. Reson. Med. **56**(2), 364–380 (2006)
11. Jezzard, P., Balaban, R.S.: Correction for geometric distortion in echo planar images from B0 field variations. Magn. Reson. Med. **34**(1), 65–73 (1995)
12. Morgan, P.S., Bowtell, R.W., McIntyre, D.J., Worthington, B.S.: Correction of spatial distortion in EPI due to inhomogeneous static magnetic fields using the reversed gradient method. J. Magn. Reson. Imaging **19**(4), 499–507 (2004)
13. Chang, H., Fitzpatrick, J.M.: A technique for accurate magnetic resonance imaging in the presence of field inhomogeneities. IEEE Trans. Med. Imaging **11**(3), 319–29 (1992)
14. Fu Y., Lei Y., Wang T., Curran W.J., Liu T., Yang X.: Deep learning in medical image registration: a review. Phys. Med. Biol. **65**, 20TR01 (2020)
15. Haskins, G., Kruger, U., Yan, P.: Deep learning in medical image registration: a survey. Mach. Vis. Appl. **31**, 8 (2020)
16. Balakrishnan, G., Zhao, A., Sabuncu, M.R., Guttag, J.V., Dalca, A.V.: VoxelMorph: a learning framework for deformable medical image registration. IEEE Trans. Med. Imaging **38**, 1788–1800 (2019)
17. Dalca, A.V., Balakrishnan, G., Guttag, J.V., Sabuncu, M.R.: Unsupervised learning of probabilistic diffeomorphic registration for images and surfaces. Med. Image Anal. **57**, 226–236 (2019)
18. Ronneberger, O., Fischer, P., Brox, T.: U-net: convolutional networks for biomedical image segmentation. In: Navab, N., Hornegger, J., Wells, W.M., Frangi, A.F. (eds.) MICCAI 2015. LNCS, vol. 9351, pp. 234–241. Springer, Cham (2015). https://doi.org/10.1007/978-3-319-24574-4_28

The Impact of Susceptibility Distortion Correction Protocols on Adolescent Diffusion MRI Measures

Talia M. Nir, Julio E. Villalón-Reina, Paul M. Thompson, and Neda Jahanshad(✉)

Imaging Genetics Center, Mark and Mary Stevens Neuroimaging and Informatics Institute, Keck School of Medicine, University of Southern California, Los Angeles, CA, USA
njahansh@usc.edu

Abstract. Diffusion MRI (dMRI) is widely used to chart the development of brain white matter (WM) microstructure across the lifespan, but suffers from susceptibility distortions and signal loss in brain regions at air-tissue boundaries like the brain stem, and ventral regions of the temporal and frontal lobes. Due to time limitations when acquiring data in adolescent, aging, and clinical populations, acquiring dMRI data twice with opposite phase encoding directions (blip-up, blip-down), as required by existing susceptibility correction tools, may not be feasible. Here we used 3T dMRI data from 99 healthy adolescents (age range: 8–21 yrs; 48% Male) from the HCP Development cohort to compare six preprocessing schemes—using either no, full, or various degrees of duplicate blip-up blip-down data as input for FSL's widely used *topup* and *eddy* distortion correction tools—to provide guidance on dMRI acquisition protocols when scan time is limited and a trade-off needs to be made. For each preprocessing pipeline, we compared the error in regional WM DTI and NODDI model fits, as well as regional associations with age. We found that model fitting errors were significantly higher in pipelines that did not use the full blip-up blip-down acquisition; associations with age were largely not affected by the preprocessing scheme used to correct susceptibility distortions.

Keywords: Susceptibility distortion · Development · White matter · Diffusion MRI

1 Introduction

Diffusion MRI (dMRI) is widely used to study brain microstructure and chart the development of the brain's white matter (WM) across the lifespan. As children transition into adolescence, their brains undergo a complex sequence of pruning and integration, making accurate modeling of WM microstructure particularly

T. M. Nir and J. E. Villalón-Reina—Contributed equally to this work.

S. Cetin-Karayumak et al. (Eds.): CDMRI 2022, LNCS 13722, pp. 50–61, 2022.
https://doi.org/10.1007/978-3-031-21206-2_5

important [1]. dMRI, however, is highly prone to a number of imaging artifacts. dMRI acquired using echo-planar imaging (EPI) is very sensitive to magnetic field inhomogeneities which can result in distortions in brain regions with extensive magnetic susceptibility variations; this can lead to distorted anatomy and signal loss in regions of the brain near air-tissue or air-bone boundaries like the brain stem, and ventral areas of the temporal and frontal lobes [2]. These distortions happen along the phase encoding (PE) direction and may either cause the signal of several voxels to "pile up" into one voxel (compression) or "smear" the signal of one voxel over several voxels (spread).

Existing susceptibility distortion correction tools, such as FSL's *topup* and *eddy* [3–5], TORTOISE's *dr-buddi* [6], and ACID's *hysco* [7] require that dMRI data be acquired, at least partially, in duplicate, with reverse PE directions (i.e., blip-up, blip-down). Due to the variety of biological, neuropsychiatric, and imaging data acquired for adolescent, aging, and clinical population studies, time constraints are often placed on imaging protocols to reduce attrition or motion and ensure adequate sample sizes. This precludes the acquisition of the longer dMRI protocols needed to optimally use existing correction tools.

The Human Connectome Project (HCP) [8,9] studies have allotted over 20 min of scan time to acquiring a full stack of dMRI volumes in both PE directions. By contrast, many clinical datasets, including ADNI [10] and PPMI [11], need to limit dMRI scan times to under 10 min and do not acquire any opposing PE volumes. Still other population studies, like ABCD [12] and UK Biobank [13], limit reverse PE to the b_0 volumes. As of yet, there is not a general consensus on the optimal acquisition for studies that are not able to acquire a full stack of b_0 and diffusion weighted images (DWI) with reversed PE.

Recently, Synb0-DISCO [14] was introduced to synthesize undistorted b_0 volumes in lieu of acquiring b_0 volumes in both PE directions. The efficacy of this tool, which was trained on one dataset (Baltimore Longitudinal Study of Aging study [15]) and validated on 26 subjects, has not yet been well established on a wide range of heterogeneous datasets (with diverse clinical, demographic and imaging quality). In addition to b_0 volumes, the question remains as to whether collecting only a small subset of DWI volumes with opposite PE directions mitigates some of the distortion and signal loss.

Here, we set out to compare measures from two dMRI models—(1) single-shell diffusion tensor imaging (DTI) [16] and (2) multi-shell neurite orientation dispersion and density imaging (NODDI) [17]—derived from data preprocessed with 1) FSL topup field estimation using either Synb0-DISCO or blip-up blip-down b_0; and 2) FSL eddy distortion correction using all b_0 but 0%, 25%, 50% or 100% of DWI volumes with reversed PE. We used dMRI data from healthy adolescents (age range: 8–21 years) from the Lifespan Human Connectome Project Development (HCP-D) cohort, to compare the effects of each preprocessing scheme on 1) the error in each dMRI model's fit and 2) regional WM microstructural associations with age. We aim to help establish guidelines about minimum dMRI acquisition protocol requirements when scan time is limited.

2 Methods

2.1 Study Participants and MRI Data

We analyzed data from the HCP-D study that includes healthy children, adolescents, and young adults, ages 8–21 years [8,9]. Here, we analyzed a subset of 99 participants spanning the same age range as the full sample (mean age: 13.9 ± 3.9 yrs.; 51 female). The HCP-D dMRI protocol was acquired on 3T Siemens Prisma scanners and includes 185 directions split across 2 shells of $b = 1500$ and $3000\,\mathrm{s/mm}^2$, and 28 $b = 0\,\mathrm{s/mm}^2$ images (multi-band factor = 4; TE/TR = 0.0892/3.23 s; 1.5 mm isotropic voxels); these were each acquired twice with reversed PE directions (i.e., 213 anterior-posterior -AP- and 213 posterior-anterior -PA- volumes). The total dMRI scan time was 22.7 min.

2.2 dMRI Preprocessing and Subsampling

Raw DWI were first denoised using DIPY's MP-PCA filter [18,19] and DIPY's Gibbs ringing suppression tool [20]. As most studies collecting dMRI data cannot collect the extensive dMRI protocol acquired in HCP-D, we used the framework presented in Zhan et al. (2010) [21] to first subsample the original 213 AP and PA volumes to 5 b_0, 50 $b = 1500\,\mathrm{s/mm}^2$, and 50 $b = 3000\,\mathrm{s/mm}^2$ volumes to match the angular resolution of the frequently used UK Biobank dMRI data [13]. From there, the AP DWI volumes from each of the two shells were subsampled to include either 50% (N = 25) or 25% (N = 13) of all the DWI volumes. Gradient subsets were selected by optimizing the spherical angular distribution energy. Briefly, the angular distribution energy, E_{ij}, of a pair of points, i and j, on the unit sphere may be defined as the inverse of the sum of the squares of (1) the least spherical distance between point i and point j, and (2) the least spherical distance between point i and point j's antipodal, symmetric point, J:

$$E_{i,j} = \frac{1}{dist^2(i,j) + dist^2(i,J)} \tag{1}$$

The total angular distribution energy $E_L(N)$ for a subset of N DWI volumes is defined as the sum of the angular distribution energy of all pairs of unit gradient vectors on the sphere, using least spherical distances. The optimal sampled gradient subset N can be achieved by selecting the set that maximizes $E_L(N)$:

$$E_L(N) = \sum\nolimits_{i=1}^{N} \sum\nolimits_{j=i+1}^{N} E_{i,j} \tag{2}$$

Susceptibility-induced off-resonance field estimation using FSL's *topup* [4,5] was then either (1) not done at all, (2) completed using a synthetic undistorted b_0 created using Synb0-DISCO [14], or (3) completed using 5 PA and 5 AP b_0s. Eddy correction was performed with FSL's *eddy* tool [3] including *repol* outlier replacement [22] with either (1) no *topup* field ('no AP'), (2) *topup* field estimation using Synb0-DISCO's synthetic b_0 ('SynB0'), (3) *topup* field estimation using 5 PA and 5 AP b_0 and no additional AP DWI volumes ('B0 AP'), (4) *topup*

Table 1. Geometric distortion correction schemes using FSL's *topup* and *eddy*.

Correction scheme	*Topup* field estimation	No. of *Eddy* AP volumes
'no AP'	None	0% AP DWI
'SynB0'	Synthetic b_0	0% AP DWI
'B0 AP'	5 PA and 5 AP b_0	0% AP DWI
'B0+25% AP DWI'	5 PA and 5 AP b_0	25% AP DWI
'B0+50% AP DWI'	5 PA and 5 AP b_0	50% AP DWI
'B0+100% AP DWI'	5 PA and 5 AP b_0	100% AP DWI

fields estimated using 5 PA and 5 AP b_0 and either 25% of the AP DWI volumes ('B0+25% AP DWI'), (5) 50% of the AP DWI volumes ('B0+50% AP DWI'), or (6) all of the AP volumes ('B0+100% AP DWI'). The resulting corrected dMRI from these six eddy processing pipelines were then skull-stripped using FSL's *bet* [23] and run through FSL's *fast* bias field inhomogeneity correction [24]. For reference, the different correction schemes can be seen in Table 1.

2.3 dMRI Models and Regional Measures

For each of the six processing schemes, DTI [16] fractional anisotropy (FA) and mean diffusivity (MD) maps were calculated with weighted least squares (FSL's *dtifit*), using only the subset of 5 b_0 and 50 $b = 1500\,\mathrm{s/mm}^2$ DWI volumes. NODDI was fitted with DMIPY [17,25], yielding maps of the intra-cellular volume fraction (ICVF) and the isotropic volume fraction (ISOVF).

To extract WM atlas ROI summary measures, first, an ANTs [26] two-channel non-linear registration was used to warp each participant's dMRI to the FSL-HCP1065 DTI template [27]. DTI FA and MD equally drove registrations to the template FA and MD. The resulting deformations were then applied to all NODDI and DTI maps. Mean dMRI measures were extracted from seven WM ROIs from the JHU stereotaxic WM atlas (Table 2) [28]. We evaluated five WM ROIs in regions of the brainstem (CP, MCP, ML) and temporal lobes (UNC, CGH) - regions that are more likely to suffer from susceptibility distortions. We also included two less distortion-prone ROIs (SLF and FullWM), which have demonstrated robust WM maturation trends in prior studies [29]. In addition to average regional NODDI and DTI scalar measures, we extracted the average mean squared error (MSE) between measured and predicted DWI signals for both DTI and NODDI models.

Table 2. Index of 7 JHU atlas WM ROIs analyzed.

MCP	Middle Cerebellar Peduncle	UNC	Uncinate Fasciculus
ML	Medial Lemniscus	CGH	Cingulum of the Hippocampus
CP	Cerebral Peduncle	Full WM	Full White Matter
SFO	Superior Fronto-Occipital Fasc		

2.4 Statistics

First, for each subject, we compared the normalized mutual information between NODDI ICVF maps generated from each of the six processing schemes to ICVF derived from 'B0+100% AP DWI' volumes (i.e., our silver standard), using the R 'aricode' package. We then tested for differences in regional DTI and NODDI model MSE values across the six processing schemes with a one-way analysis of variance (ANOVA).

As WM development trajectories are not entirely linear, we used generalized additive models (GAMs) to test for associations between age and mean dMRI measures in each ROI using the R 'mgcv' package. Age and age-by-sex interactions were modeled using spline smoothing functions (cubic regression b-splines, k = 10); sex and acquisition site were included as fixed effect covariates. The false discovery rate (FDR) procedure was used to correct for multiple comparisons across ROIs ($q = 0.05$) [30]. Differences in resulting age F-values and model adjusted R^2 across the six processing schemes were tested with ANOVA.

3 Results

3.1 DTI and NODDI Map Comparisons

As illustrated in Fig. 1, qualitatively, ICVF map distortions from an example subject are gradually less visible (Fig. 1A) and normalized mutual information between subjects' ICVF maps derived from 'B0+100% AP DWI' volumes (i.e., our silver standard) and each processing scheme increases (Fig. 1B) as the number of AP DWI volumes included in the FSL eddy corrections increases. Whole brain normalized correlations between DTI and NODDI maps derived from each processing scheme confirm greater similarities when any number of AP DWI are used (25%, 50%, or 100%; Fig. 2).

3.2 DTI and NODDI Fit Evaluations

MSE in DTI and NODDI model fits gradually decreased as the number of AP DWI volumes included in the FSL eddy corrections increased with the most signal recovered in the temporal lobes with 'B0+100% AP DWI' volumes (Fig. 1C,D). ANOVA revealed significant MSE group differences between processing schemes across all seven ROIs in both dMRI models (Fig. 3). Post-hoc pairwise t-tests were subsequently performed to directly compare ROI MSEs between each pair of pipelines and corrected with FDR for 15 pairwise tests. We found fewer significant differences in regional DTI or NODDI MSE between 'no AP', 'SynB0', and 'B0 AP', while '50% AP' and '100% AP' almost always showed stepwise improvements (i.e., decreases) in MSE.

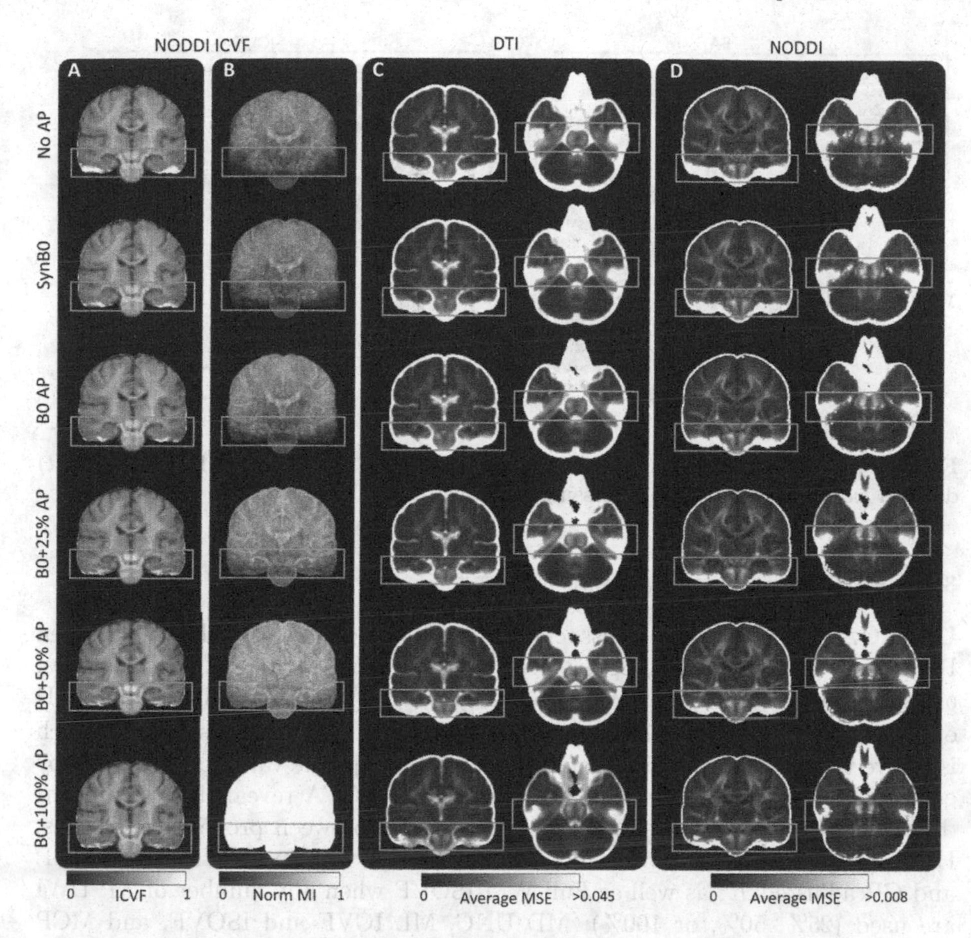

Fig. 1. (A) Examples of NODDI ICVF maps are shown for a single subject. (B) Normalized mutual information between subjects' 'B0+100% AP DWI' derived ICVF and ICVF derived from each of the six processing pipelines. (C) Mean squared error (MSE) between measured and predicted DWI signals for DTI and (D) NODDI models, averaged across all participants. Temporal regions prone to susceptibility distortions are highlighted with orange boxes. (Color figure online)

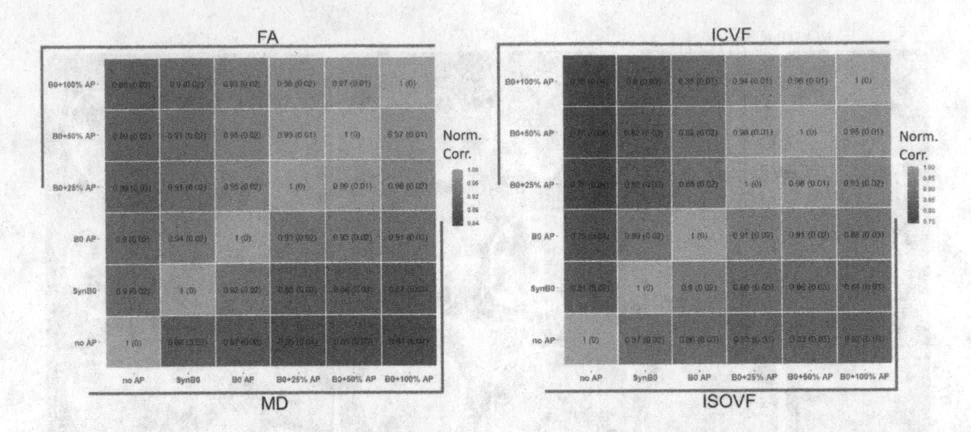

Fig. 2. Normalized correlations between full brain DTI (*left*) or NODDI maps (*right*) derived from each of the six processing pipelines.

3.3 GAM Age Associations

Using GAMs, significant age associations were detected across almost all seven ROIs (i.e., CP, MCP, ML, UNC, CGH, SLF, Full WM), four dMRI measures, and six processing schemes evaluated (Fig. 4). We note that DTI FA and MD effects may be influenced in part by the high b-value ($b = 1500\,\mathrm{s/mm^2}$), which is above the optimal diffusion weighting for DTI [30]. The variance explained by each model (adjusted R^2) is shown in Fig. 4B. ANOVA revealed no significant differences in age F-values or model adjusted R^2 between processing schemes. However, qualitatively, improvements can be seen in MD and ICVF for the MCP and CP adjusted R^2 as well as Full WM ISOVF when any number of AP DWI are used (25%, 50%, or 100%); MD UNC, ML ICVF and ISOVF, and MCP ISOVF adjusted R^2 only improve when using 100% AP DWI. For reference, MD and ICVF values in the CP and UNC are plotted in Fig. 5.

4 Discussion

Overall, we found that while DTI and NODDI errors in fit were significantly higher in pre-processing pipelines that did not use the full set of blip-up blip-down DWI volumes, associations with age were largely unaffected. While we hypothesized that the preprocessing schemes using a greater number of AP volumes would perform significantly better, it is possible that age effects in these regions are too robust to be affected by small but significant differences in model fit. Future work examining either microstructure in more peripheral WM regions or cortical gray matter, and evaluating more complex microstructural relationships with, for example, genetics or cognition, may reveal different trends.

Another factor that may drive similarities between methods is registration; as in the EPI distortion correction method whereby each subject's b_0 is warped

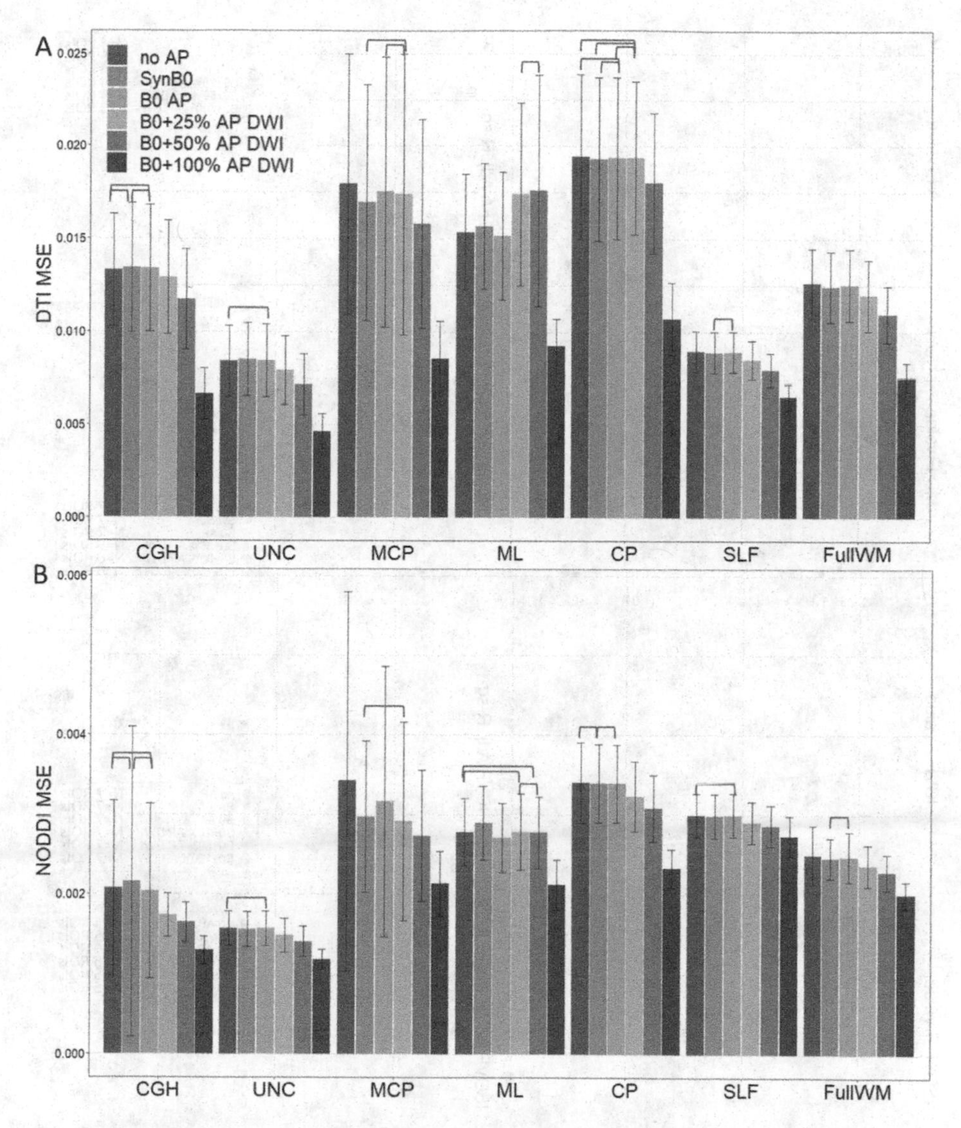

Fig. 3. Average regional mean squared error (MSE) between measured and predicted DWI signals for (A) DTI (single-shell; *top*) and (B) NODDI (multi-shell; *bottom*). Given that pairwise tests between processing schemes revealed that the majority of the MSE differences were significant, differences that are **NOT** significant (FDR corrected $p > 0.05$) are demarcated by red brackets. (Color figure online)

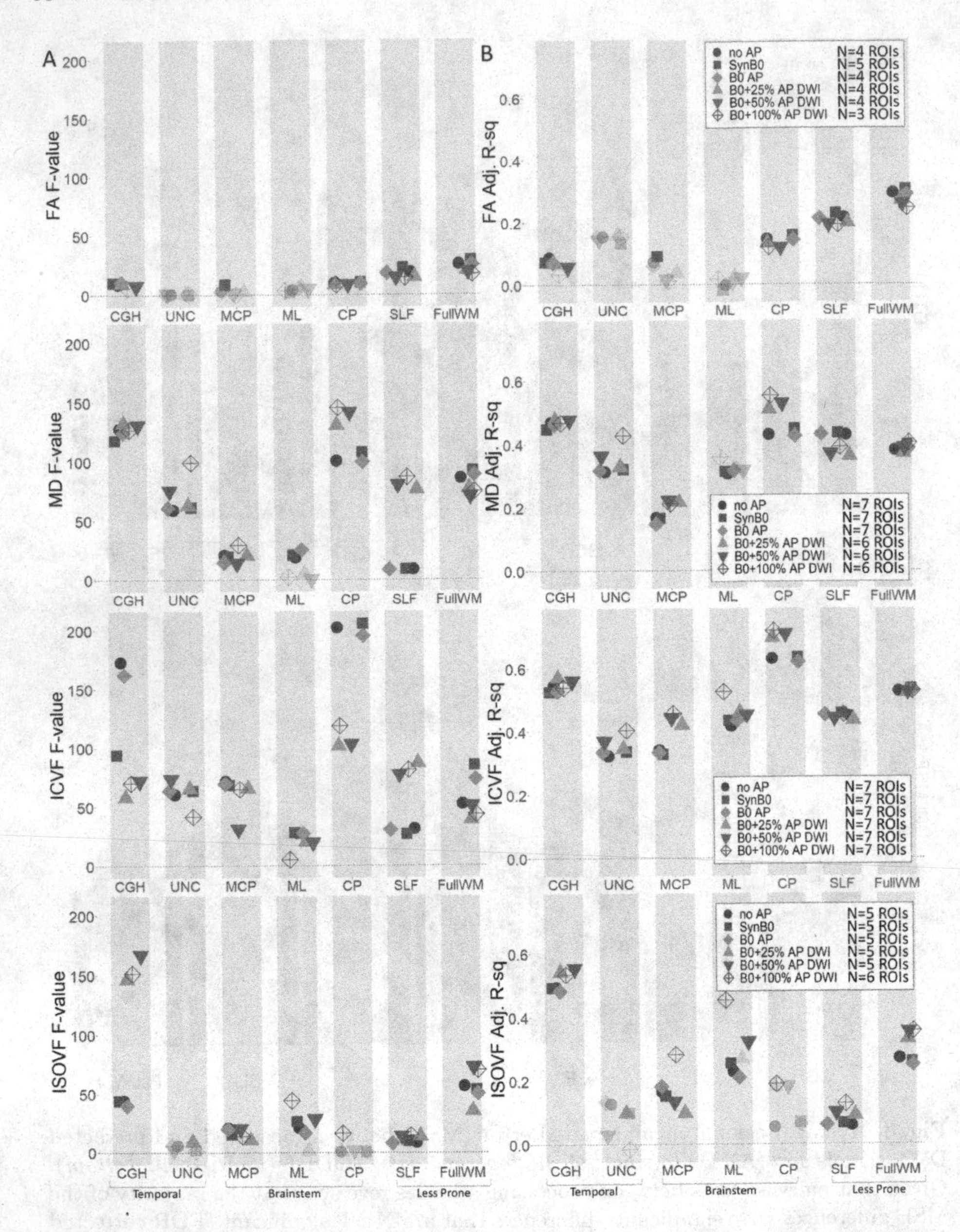

Fig. 4. (A) F-values (*left*) and (B) adjusted R^2 (*right*) for GAM models evaluating associations between age and regional DTI FA and MD (single-shell) or NODDI ICVF and ISOVF (multi-shell). The number of ROIs that were significant after multiple comparisons correction (FDR corrected $p < 0.05$) is noted in the legend as 'N'; significant ROIs are shown as opaque in the figure.

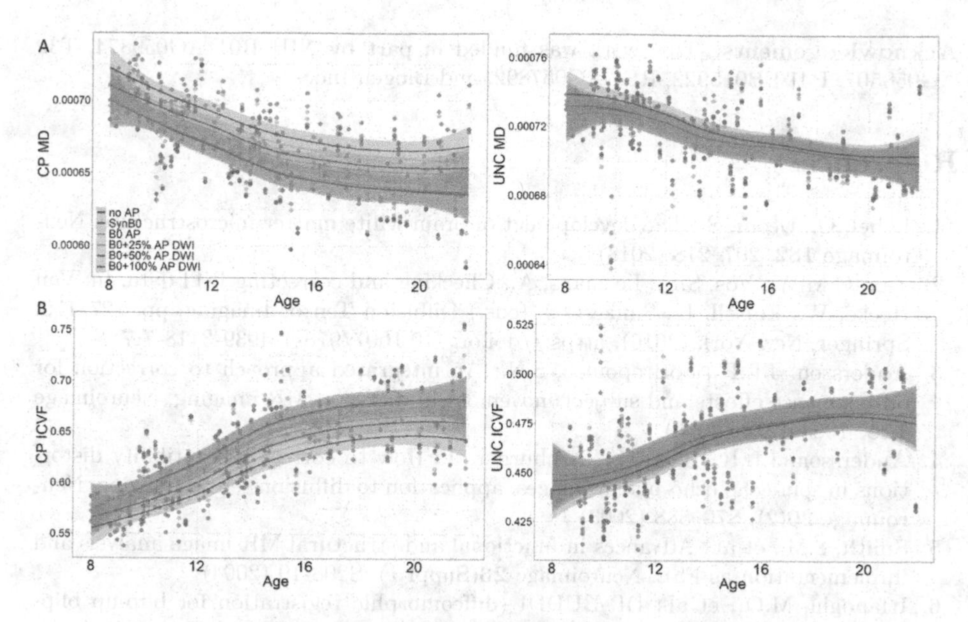

Fig. 5. MD and ICVF in the (A) CP (*top*) and (B) UNC (*bottom*), plotted against age. For the CP, lower average MD for correction schemes with no AP DWIs ('no AP', 'SynB0', 'B0 AP') is evident; MD gradually increases the more AP DWIs were included in FSL *eddy* (25%, 50%, 100% AP DWI). The reverse effect is evident for ICVF. This trend is more subtle for the UNC.

to their respective undistorted T1-weighted anatomical image, the registration of subject dMRI maps to the template may have helped mitigate some remaining distortions across approaches. However, nonlinear registrations are dependent on accurate brain masks - which may be difficult to automatically delineate in distorted images. Such issues may be alleviated by using alternative methods that are more robust to misregistration, such as TBSS [31], to extract regional diffusion measures.

In addition, we found that processing schemes that included no reversed PE DWI volumes (i.e., 'no AP', 'SynB0', and 'B0 AP') showed similar MSE to one another (i.e. fewer significant differences). This suggests that when scan time is very limited, there might be little benefit in dMRI protocols that only acquire b_0 volumes with reversed PE; it might be more beneficial to focus on optimizing angular or spatial resolution instead. The lack of significant improvements in MSE when using the true 'B0 AP' could be driven in small part by relatively greater levels of noise and artifacts in true b_0 data compared to synthetic or no b_0.

Future studies should examine differences in independent cohorts to establish the degree to which these results can be generalized to other age ranges and datasets, in addition to evaluating the sensitivity of biological variables of interest beyond age. These studies will help further establish minimum dMRI acquisition protocol requirements needed to meet specific research goals when scan time is limited and a trade-off needs to be made.

Acknowledgements. This work was funded in part by NIH R01 AG059874, T32 AG058507, P41 EB015922, RF1 AG057892, and Biogen Inc.

References

1. Lebel, C., Deoni, S.: The development of brain white matter microstructure. Neuroimage **182**, 207–218 (2018)
2. Tax, C.M.W., Vos, S.B., Leemans, A.: Checking and correcting DTI data. In: Van Hecke, W., Emsell, L., Sunaert, S. (eds.) Diffusion Tensor Imaging, pp. 127–150. Springer, New York (2016). https://doi.org/10.1007/978-1-4939-3118-7_7
3. Andersson, J.L.R., Sotiropoulos, S.N.: An integrated approach to correction for off-resonance effects and subject movement in diffusion MR imaging. Neuroimage **125**, 1063–1078 (2016)
4. Andersson, J.L.R., Skare, S., Ashburner, J.: How to correct susceptibility distortions in spin-echo echo-planar images: application to diffusion tensor imaging. Neuroimage **20**(2), 870–888 (2003)
5. Smith, S.M., et al.: Advances in functional and structural MR image analysis and implementation as FSL. Neuroimage **23**(Suppl 1), S208-19 (2004)
6. Irfanoglu, M.O., et al.: DR-BUDDI (diffeomorphic registration for blip-up blip-down diffusion imaging) method for correcting echo planar imaging distortions. Neuroimage **106**, 284–299 (2015)
7. Ruthotto, L., et al.: Diffeomorphic susceptibility artifact correction of diffusion weighted magnetic resonance images. Phys. Med. Biol. **57**(18), 5715–5731 (2012)
8. Harms, M.P., et al.: Extending the human connectome project across ages: imaging protocols for the lifespan development and aging projects. Neuroimage **183**, 972–984 (2018)
9. Somerville, L.H., et al.: The lifespan human connectome project in development: a large-scale study of brain connectivity development in 5–21 year olds. Neuroimage **183**, 456–468 (2018)
10. Zavaliangos-Petropulu, A., et al.: Diffusion MRI indices and their relation to cognitive impairment in brain aging: the updated multi-protocol Approach in ADNI3. Front. Neuroinform. **13**, 2 (2019)
11. Initiative, P.P.M.: The Parkinson progression marker initiative (PPMI). Prog. Neurobiol. **95**(4), 629–635 (2011)
12. Hagler, D.J., Jr., et al.: Image processing and analysis methods for the adolescent brain cognitive development study. Neuroimage **202**, 116091 (2019)
13. Miller, K.L., et al.: Multimodal population brain imaging in the UK Biobank prospective epidemiological study. Nat. Neurosci. **19**(11), 1523–1536 (2016)
14. Schilling, K.G., et al.: Synthesized b0 for diffusion distortion correction (Synb0-DisCo). Magn. Reson. Imaging **64**, 62–70 (2019)
15. Ferrucci, L.: The Baltimore Longitudinal Study of Aging (BLSA): a 50-year-long journey and plans for the future. J. Gerontol. A Biol. Sci. Med. Sci. **63**(12), 1416–1419 (2008)
16. Basser, P.J., Mattiello, J., LeBihan, D.: MR diffusion tensor spectroscopy and imaging. Biophys. J . **66**(1), 259–267 (1994)
17. Zhang, H., et al.: NODDI: practical in vivo neurite orientation dispersion and density imaging of the human brain. Neuroimage **61**(4), 1000–1016 (2012)
18. Veraart, J., et al.: Denoising of diffusion MRI using random matrix theory. Neuroimage **142**, 394–406 (2016)

19. Garyfallidis, E., et al.: Dipy, a library for the analysis of diffusion MRI data. Front. Neuroinform. **8**, 8 (2014)
20. Veraart, J., et al.: Gibbs ringing in diffusion MRI. Magn. Reson. Med. **76**(1), 301–314 (2016)
21. Zhan, L., et al.: How does angular resolution affect diffusion imaging measures? Neuroimage **49**(2), 1357–1371 (2010)
22. Andersson, J.L.R., et al.: Incorporating outlier detection and replacement into a non-parametric framework for movement and distortion correction of diffusion MR images. Neuroimage **141**, 556–572 (2016)
23. Smith, S.M.: Fast robust automated brain extraction. Hum. Brain Mapp. **17**(3), 143–155 (2002)
24. Zhang, Y., Brady, M., Smith, S.: Segmentation of brain MR images through a hidden Markov random field model and the expectation-maximization algorithm. IEEE Trans. Med. Imaging **20**(1), 45–57 (2001)
25. Fick, R.H.J., Wassermann, D., Deriche, R.: The Dmipy toolbox: diffusion MRI multi-compartment modeling and microstructure recovery made easy. Front. Neuroinform. **13**, 64 (2019)
26. Avants, B.B., et al.: A reproducible evaluation of ANTs similarity metric performance in brain image registration. Neuroimage **54**(3), 2033–2044 (2011)
27. Yeh, F.-C., et al.: Population-averaged atlas of the macroscale human structural connectome and its network topology. Neuroimage **178**, 57–68 (2018)
28. Mori, S., et al.: Stereotaxic white matter atlas based on diffusion tensor imaging in an ICBM template. Neuroimage **40**(2), 570–582 (2008)
29. Lebel, C., Beaulieu, C.: Longitudinal development of human brain wiring continues from childhood into adulthood. J. Neurosci. **31**(30), 10937–10947 (2011)
30. Benjamini, Y., Hochberg, Y.: Controlling the false discovery rate: a practical and powerful approach to multiple testing. J. R. Stat. Soc. **57**(1), 289–300 (1995)
31. Smith, S.M., et al.: Tract-based spatial statistics: voxelwise analysis of multisubject diffusion data. Neuroimage **31**(4), 1487–1505 (2006)

Signal Representations

Diffusion MRI Fibre Orientation Distribution Inpainting

Zihao Tang[1,2](✉), Xinyi Wang[1,2], Mariano Cabezas[2], Arkiev D'Souza[2], Fernando Calamante[2], Dongnan Liu[1,2], Michael Barnett[2,3], Chenyu Wang[2,3], and Weidong Cai[1]

[1] School of Computer Science, University of Sydney, Sydney, NSW 2008, Australia
ztan1463@uni.sydney.edu.au
[2] Brain and Mind Centre, University of Sydney, Sydney, NSW 2050, Australia
[3] Sydney Neuroimaging Analysis Centre, Sydney, NSW 2050, Australia

Abstract. The analysis of diffusion weighted brain magnetic resonance images, including the estimation of fibre orientation distribution (FOD), tractography, and connectomics, is a powerful tool for neuroscience research and clinical applications. However, focal brain pathology and imaging acquisition artifacts affecting white matter tracts may disrupt or corrupt FOD values respectively, invalidating tractography and connectome reconstructions. In this work, we propose a 3D FOD inpainting framework, named order-wise coefficient estimation network (OCE-Net), to dynamically reconstruct the affected regions. Our feature encoding stage, based on gated convolutions, extracts features from all the input FOD coefficients and re-weights them using channel attention and independent order-wise decoders, to independently predict the coefficients for each spherical harmonic order. We evaluated our model on a subset of scans from the HCP dataset, and conducted tractography and connectomics to further analyse the impact of inpainting. Our experimental results, including a statistical analysis of the reconstructed connectomes, show that our OCE-Net approach can successfully reconstruct the original FODs in the focally disrupted regions.

Keywords: Diffusion MRI · Fibre orientation distribution · Inpainting · 3D gate network

1 Introduction

Diffusion weighted imaging (DWI) is a magnetic resonance imaging (MRI) technique that can be used to visualise white-matter fibre bundles in vivo. It serves as a powerful tool for studying the organisation and microstructure of white-matter in healthy and diseased populations; and has broad applications in neuroscience research and clinical translation, where it may improve diagnosis, facilitate precision monitoring, and guide surgical intervention for a variety of brain diseases.

Z. Tang and X. Wang—Equal contributions.

S. Cetin-Karayumak et al. (Eds.): CDMRI 2022, LNCS 13722, pp. 65–76, 2022.
https://doi.org/10.1007/978-3-031-21206-2_6

Derivatives of DWI include whole-brain tractography and connectomics [29], which can be used to investigate global changes in brain connectivity (e.g., using graph metrics), and probe the connectivity of specific regions of interest (ROIs) with the rest of the brain. A key intermediate step for tractography is the computation of the white-matter fibre orientation distribution (FOD) image [25]. This image is typically generated by applying constrained spherical deconvolution to the pre-processed DWI data [8,11,24]. In the context of tractography, the FOD models local fibre direction(s), and permits the reconstruction of crossing and intersecting fibres.

Failure to generate accurate FODs can have deleterious repercussions for tractography. Absent or disrupted FODs could lead to inaccurate tractograms [6], and subsequently, incorrect connectivity and graph theory estimates. Erroneous FODs may arise from focal, destructive brain pathology [20] or changes in diffusivity due to compromised white-matter integrity [12]. Such focal disruptions may also subsequently prevent the subject FODs from being accurately warped to the population template, and may result in the image being discarded. These scenarios can be particularly problematic in neurological conditions such as multiple sclerosis (MS), where there is substantial interest in quantifying white-matter integrity in the presence of focal, destructive lesions [13]. While traditional intensity-based neuroimaging analysis pipelines apply lesion inpainting methods [20] on the corresponding T1 image to prevent the mis-classification of whiter matter lesions as gray matter [21], diffusion image analysis is critically dependent on the FOD. There is therefore incentive to improve the quality of the FODs in ROIs disrupted by focal pathology (or also by acquisition artefacts).

To this end, we propose an order-wise coefficient estimation network (OCE-Net)[1] for restoring disrupted ROIs through inpainting, based on the neighbouring voxels. OCE-Net consists of a feature encoding stage and an order-wise coefficient estimation stage, which are specifically designed for inpainting FODs. To the best of our knowledge, this is the first work that addresses inpainting directly on the FODs, rather than the raw DWI sequence. Our evaluation results show that the proposed OCE-Net is capable of recovering the original FODs on a set of images from the Human Connectome Project (HCP) with any localized FOD disruption. Furthermore, downstream tasks such as the connectome construction, and graph-based connectivity metrics presented no significant differences when the results derived from inpainted images and the original FOD images were compared.

2 Methods

2.1 The Human Connectome Project Dataset

50 different subjects were selected from the Human Connectome Project (HCP) database[2] [28]. The HCP dataset was acquired using a 3T Siemens 'Connectom'

[1] The project page of this work is available at: https://mri-synthesis.github.io.
[2] https://www.humanconnectome.org/.

Skyra scanner, and included anatomical T1-weighted imaging and diffusion-weighted imaging. The high resolution T1-weighted data were acquired with 0.7 mm isotropic resolution, TR/TE = 2400/2.14 ms, and flip angle = 8°. The diffusion MRI protocol consisted of three diffusion-weighted shells (b=1000, 2000, and 3000 s/mm^2, respectively, with 90 directions per shell) and eighteen reference volumes. The diffusion MRI data were acquired with 1.25 mm isotropic resolution, TR/TE = 5520/89.5 ms, and flip angle = 78°.

2.2 Data Preprocessing

The preprocessing steps for raw diffusion images included corrections for motion, susceptibility distortions, gradient nonlinearity and eddy currents [1,2,19]. Bias correction was then applied to the structural images [27], followed by a multi-shell multi-tissue constrained spherical deconvolution to generate the FODs in each voxel [11,26]. These steps were performed consistently on each subject using the MRtrix software package v3.0.2 [22,23]. The resulting FOD images had a resolution of 145 × 174 × 145 × 45, 45 being the number of spherical harmonic coefficients (corresponding to lmax = 8).

To analyze the effect of inpainting on disrupted ROIs in a controlled environment, we generated a synthetic mask for each subject based on the lesion distribution of a set of MS patients as described in [21]. To simulate the influence of the disrupted ROIs in real-world cases, Gaussian noise was added in the masked ROIs and the resulting spherical harmonics coefficients of FOD were then multiplied by an attenuation coefficient of 0.5 to simulate the effect of signal loss.

To further study the effect of disrupted and inpainted FODs on downstream tasks, and therefore investigate their practical consequences, we performed tractography and constructed connectomes with the following pipeline. Five tissue type (5TT) segmentations were first generated by hybrid surface and volume segmentation [15]. The 5TT image and the corresponding FOD data were then used to produce 10 million streamlines using anatomically constrained probabilistic tractography with dynamic seeding [16,25]. The parameters used for the tractography framework followed the default settings in MRtrix, except that the maximum length was set to 300 mm and the following options enabled: dynamic seeding, backtracking and cropping at the grey-matter-white-matter interface. A weight was assigned to each streamline by applying spherical-deconvolution informed filtering of tractograms 2 (SIFT2, [18]) on the streamlines, in order to integrate the biological information into the connectivity measurements [17]. The T1-weighted images were segmented according to the Desikan-Killiany atlas [5,9], to generate a parcellation image consisting of 84 discrete nodes. Finally, a weighted-undirected connectome matrix was generated using the 10 million streamlines (and their corresponding SIFT2 weights).

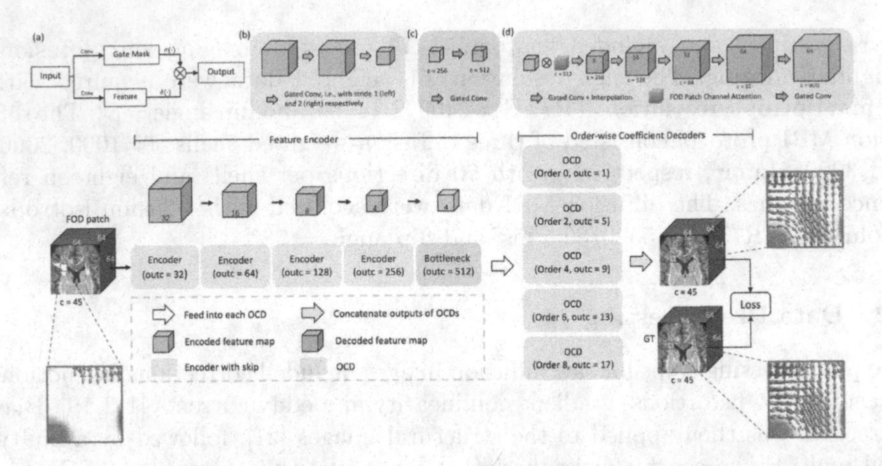

Fig. 1. The training procedure of the proposed OCE-Net. The skip connections between each block in the feature encoder and decoder are omitted for clarity. (a) Gated convolution scheme designed to compute a dynamic gate mask to guide the inpainting. (b) The architecture of the encoder block (in blue). (c) The architecture of the bottleneck between the encoding and decoding stages (in grey). (d) The architecture of the decoder for each order of the spherical harmonic coefficients (in orange). (Color figure online)

2.3 3D FOD Inpainting Framework

In this work, we present a novel 3D FOD inpainting framework, named OCE-Net, to inpaint free-form disrupted ROIs on FOD images in an end-to-end manner. Our framework, extends the encoder and decoder structure of the original 3D U-Net [4] based on the characteristics of FODs: (1) the feature encoding stage is designed to extract high-level features from the coefficients of each voxel outside the disrupted ROIs; and (2) the order-wise coefficient estimation stage estimates the final coefficients with a dynamic selection of encoded features. The details of our proposed OCE-Net are illustrated in Fig. 1. To avoid using the disrupted FOD values as the input of our framework, the disrupted ROIs under the masks were treated as empty holes. Due to memory constraints, each input volume was cropped into patches of size $64 \times 64 \times 64 \times 45$ using a sliding window of stride 32.

2.4 Feature Encoding Stage

The features from the disrupted FODs are dissimilar to those of normal-appearing regions and we argue that the true FODs can be reconstructed based solely on these neighboring regions. However, typical convolutional layers treat all voxels equally and do not account for unreliable voxel values. To address this shortcoming, we used a trainable dynamic gate mask block [30] in each 3D convolutional layer to compute attention scores in an efficient and light-weight manner [14]. Thus, our final feature encoding stage consists of four gated convolution encoders with 32, 64, 128, and 256 filters, respectively, that are defined as:

$$GatedConv = g(x) = \delta(W_{feature} \cdot x) \odot \sigma(W_{gate} \cdot x), \quad (1)$$

$$f_{encoder}(x_{in}) = g_2(g_1(x_{in})), \quad (2)$$

where $W_{feature}$ and W_{gate} are two 3D convolutional weight matrices. $W_{feature}$ extracts high-level features, while W_{gate} provides a re-weighting mechanism for all the voxels across all the channels dynamically; δ is an activation function, LeakyReLU with a slope of 0.2 in our case; σ is the sigmoid function to guarantee voxel weights in the $[0, 1]$ range; and, g_1, g_2 denote two gated convolutions with stride 1 and stride 2, respectively. Therefore, our proposed encoders can learn and highlight regions and semantic information for each channel independently at different resolution levels.

2.5 Order-Wise Coefficient Decoders

FODs encode local white matter fibre orientations as coefficients of spherical harmonics represented as a vector of fixed size (based on the order of the harmonics) despite the number of diffusion gradients used during the DWI acquisition [11,26]. The number of coefficients is commonly constrained to 45 coefficients by setting the maximum harmonic order to 8.

To reconstruct the disrupted ROIs with a proper selection of extracted features, we present an order-wise coefficient estimation stage based on the following observation of the characteristics of the FODs: lower order coefficients are known to get influenced by the higher order ones [3]. Consequently, we designed a single encoding path to extract features from all the even spherical harmonic coefficients together and analyse such dependencies.

However, to allow for flexibility in terms of maximum harmonic order selection, our proposed estimation stage comprises five order-wise coefficient decoders (OCDs) that can compute the output of each even order ([0, 2, 4, 6, 8]) independently. Compared to other conventional medical imaging techniques, FOD volumes have a high number of channels due to the number of coefficients per spherical harmonic order. To calibrate the channels for each decoder, channel attention [10] is used as the input to the OCD blocks to model the relationships between channels for each order: the spatial dimensions are squeezed using global pooling and the channels are excited and re-weighted using a fully connected layer followed by a sigmoid activation function.

Moreover, applying attention for each channel can highlight high-level spatial information in the masked regions. Therefore, our decoding stage can reconstruct the disrupted regions with the proper high-level features for each harmonic order, independently. Finally, we concatenate the outputs of each decoder to generate the final image prediction.

2.6 Implementation Details

We conducted a 5-fold cross validation on a subset of 50 subjects from the HCP dataset for all experiments. All the networks were trained for 50 epochs

Table 1. Summary of the quantitative metrics for the restoration of the disrupted FOD values. The best performance measures (the lowest MAE and MSE; and the highest PSNR) are marked in bold. w/o OCDs represents a single decoder setting and * denotes the use of channel attention.

Method	MAE ↓	MSE ↓	PSNR ↑
Empty ROIs	0.0961 ± 0.0059	0.0284 ± 0.0036	15.5077 ± 0.5500
Noisy ROIs	0.0581 ± 0.0026	0.0089 ± 0.0008	20.5109 ± 0.4111
U-Net [4]	0.0442 ± 0.0029	0.0044 ± 0.0006	23.5429 ± 0.6261
w/o OCDs	0.0424 ± 0.0028	0.0042 ± 0.0006	23.8114 ± 0.6231
w/o OCDs *	0.0421 ± 0.0028	0.0041 ± 0.0006	23.8962 ± 0.6210
OCE-Net	**0.0403 ± 0.0027**	**0.0038 ± 0.0005**	**24.1856 ± 0.6260**

using the L_1 loss for each harmonic order and the Adam optimizer with an initial learning rate of 0.001 and minibatches of 4 image patches on a single NVIDIA GeForce Tesla V100-SXM2 GPU (PyTorch 1.10.1 and Python 3.7.11). Spherical harmonic coefficients of FOD images were z-score normalized before being cropped into patches. During the training phase, we discarded any patches outside the disrupted ROIs to reduce the computational complexity of regions ignored by our loss function. During inference we reconstructed the final FOD image from the prediction of all the possible input patches inside the disrupted regions.

3 Experimental Results

3.1 Inpainting Quality Analysis

To analyse the ability of the network to restore the original FOD coefficient values, we used the following commonly used metrics to compare the predictions to the original images: mean squared error (MSE), mean absolute error (MAE), and the peak signal-to-noise ratio (PSNR). For completeness, we compared our OCE-Net framework to the original 3D U-Net [4], the current state-of-the-art for a diverse range of medical image processing tasks; and conducted several ablation studies to validate our extensions. Furthermore, we also computed all the metrics for the images with noisy and empty values (holes) within the disrupted regions. The experimental results shown in Table 1 represent the average across the five folds and demonstrate that our proposed OCE-Net achieved the best performance among all competing methods with minimum MAE and MSE, and the highest PSNR score.

In addition to traditional computer vision metrics used for reconstruction, we also included an FOD-specific metric to calculate the error of the estimated orientations of the white matter fibres (peaks) [18]. The max peak error [31] in each 'fixel' (corresponding to a fibre population within a voxel [7]) inside the disrupted ROIs was calculated for each subject and we report the mean, max, and percentage-wise max peak error for all the subjects in Table 2.

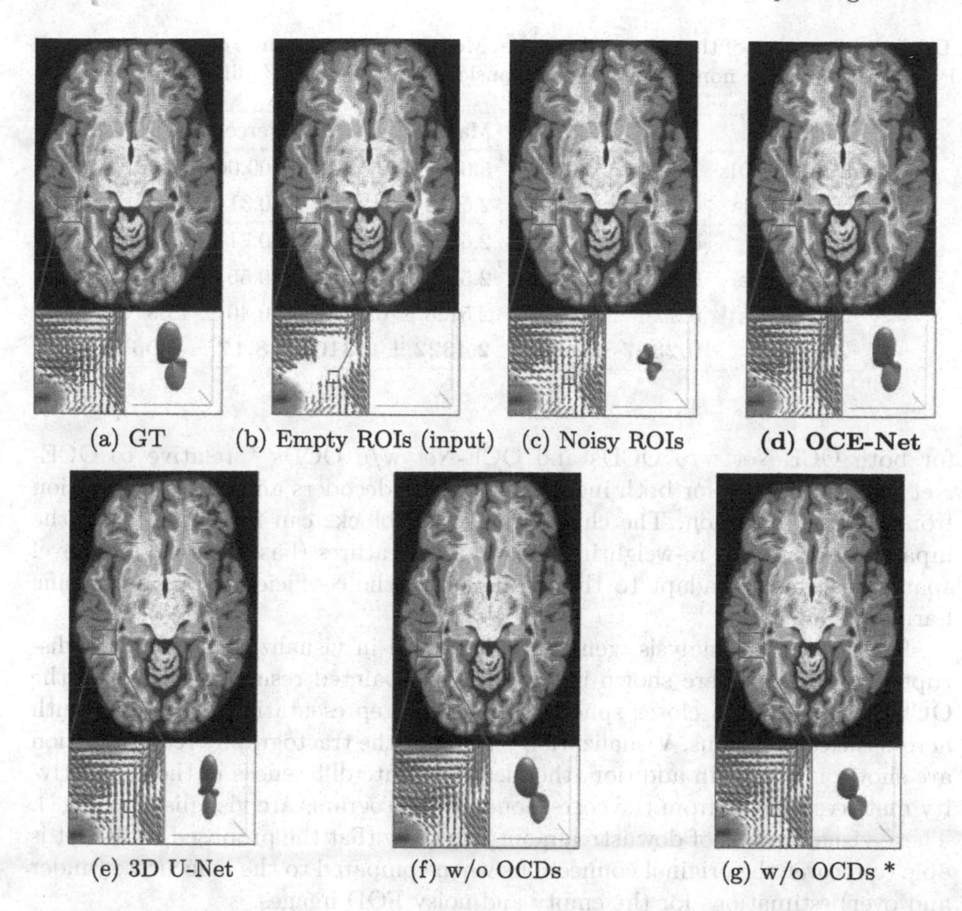

Fig. 2. Qualitative FOD inpainting results in the disrupted ROIs (red bounding box in the whole brain image) overlaid on a 5-tissue-type image. The color-coding of the sperical harmonic representation indicates directionality, whereby red, green and blue represent right-left, anterior-posterior, and inferior-superior directions, respectively. A significant example of a disrupted ROI and a FOD voxel are zoomed in for visualization: (a) ground truth, (b) empty value, (c) noisy value, inpainted value predicted by (d) OCE-Net, (e) 3D U-Net, (f) OCE-Net w/o OCDs, and (g) OCE-Net w/o OCDs *, respectively. (Color figure online)

In terms of ablation studies, we trained OCE-Net with a single decoder with gated convolutions (denoted as w/o OCDs) to determine the importance of the independent order-wise coefficient estimation. We also studied the use of channel attention with a single decoder setting (denoted as w/o OCDs *) to validate the efficacy of OCDs. According to the experimental results, OCE-Net w/o OCDs * outperforms a 3D U-Net with conventional convolutional blocks across all qualitative metrics, proving that the introduction of gated convolutions and incorporation of such an attention mechanism can extract better features for the masked regions and restore the disrupted FODs. The lower performance

Table 2. Summary of the max peak results. Mean, max and mean % error are reported. Peak values of 0 for non-zero voxels are considered to have 100% difference.

Method	Mean ↓	Max ↓	Percentage ↓
Empty ROIs	1.2959 ± 0.0774	5.0507 ± 0.5790	100.00 ± 0.00
Noisy ROIs	0.5539 ± 0.0391	2.5010 ± 0.2849	40.31 ± 0.74
U-Net	0.2665 ± 0.0360	2.6021 ± 0.4782	20.71 ± 1.95
w/o OCDs	0.2620 ± 0.0322	2.5559 ± 0.4901	20.55 ± 1.68
w/o OCDs *	0.2586 ± 0.0327	2.5128 ± 0.5037	20.40 ± 1.68
OCE-Net	**0.2307 ± 0.0253**	**2.4322 ± 0.3103**	**18.17 ± 0.96**

for both OCE-Net w/o OCDs and OCE-Net w/o OCDs * relative to OCE-Net proves the need for both independent order decoders and the re-calibration from channel attention. The channel attention blocks can further enhance the inpainting results by re-weighting the encoded features (based on the high-level spatial features) to adapt to the prediction of the coefficients for one specific harmonic order.

For qualitative analysis, general and zoomed in visualization results of disrupted FOD voxels are shown in Fig. 2. The inpainted result generated by the OCE-Net achieved a closer spherical harmonic representation to ground truth across the comparisons. Visualization results for the tractography reconstruction are shown in Fig. 3. In addition, the mean absolute differences in the connectivity matrices derived from the corresponding tractograms are visualised in Fig. 4. These visualizations of downstream outputs show that the proposed OCE-Net is able to recover the original connections when compared to the inaccurate (under and over) estimations for the empty and noisy FOD images.

3.2 Connectome Matrix Analysis

To further demonstrate the effectiveness of inpainting, we performed a battery of statistical tests between the connectomes derived from the ground truth FOD, and the disrupted FODs that had been recovered with OCE-Net. Due to the probabilistic nature of tractography, variability is introduced into the connectome construction for a given FOD image. To account for how that variability can affect the analysis, we randomly selected a training fold and constructed 10 connectome matrices per subject and image. Then, we performed multiple comparisons between the distributions of connectomes derived from processed connectomes (empty, noisy and inpainted) and the ground truth images for each connection and subject independently and counted the percentage of significantly different edges (different edges % in Table 3). Since the repeated connectome edges did not pass a normalcy test, we used the Mann-Whitney U test and defined an overall significance α of 0.01. We used Bonferroni's method to correct for multiple comparisons and set a significance value per test of $\frac{\alpha}{n}$, where $n = 3486$ was the number of unique edges in the graph, excluding the diagonal. We

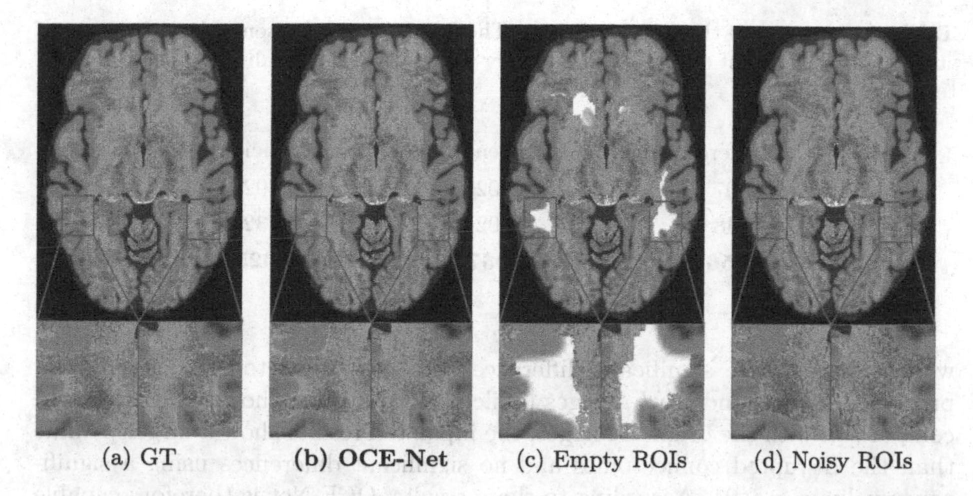

Fig. 3. Qualitative visualisations of the tractography (down-sampled to 1 million streamlines) estimates. The color-coding of the streamlines indicates directionality, whereby red, green and blue represent right-left, anterior-posterior, and inferior-superior directions, respectively. Two examples of successful reconstructions of the tractography are zoomed in for visualization: (a) ground truth, (b) inpainted values predicted by OCE-Net, (c) empty values, and (d) noisy values, respectively. (Color figure online)

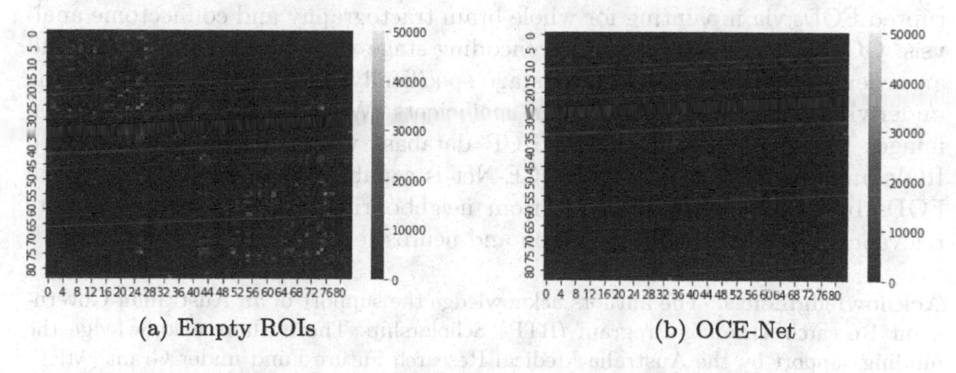

Fig. 4. Qualitative visualisations of the mean of all subject differences between the connectome matrices derived from the tractograms of the (a) empty ROIs or (b) inpainted values and the ground truth. Each node (columns and rows) represents a brain region and the edges (cells) indicate the connection between regions color-coded by the edge weight (white and black being highest and lowest, respectively).

were able to recover 98.44% of the original connections from the empty ROIs and halved the number of disrupted connections due to noise. Finally, to study the differences at an inter-subject level for the whole connectome, we computed two commonly used network metrics (strength and efficiency) per connectome as suggested in [17] and averaged the repeated values per subject to determine

Table 3. Summary of the statistical tests. The first column represents the percentage of statistically significant edges, while the other two contain the p-values for the Wilcoxon tests ($\alpha = 0.01$).

Pipeline	Different edges (%)↓	Strength (p-value) ↑	Efficiency (p-value) ↑
Empty ROIs	33.07	0.002	0.002
Noisy ROIs	2.48	0.002	0.232
OCE-Net	**1.56**	**0.0371**	**0.625**

whether there were significant differences between connectomes derived from processed and ground truth images (Wilcoxon rank tests, shown in Table 3). In comparison with the original images, the inpainted results show a larger p-value than the disrupted connectomes and no significant differences using a significance value $\alpha = 0.01$. According to these results, OCE-Net is therefore capable of recovering the original tractography, which is essential for downstream tasks and neuroscience applications.

4 Conclusion

In this work, we proposed OCE-Net, a new framework for reconstructing disrupted FODs via inpainting for whole brain tractography and connectome analysis. OCE-Net combines a feature encoding stage with gated convolutions and an order-wise coefficient decoding stage specifically designed for predicting the underlying true spherical harmonic coefficients. We evaluated our method on images from the public available HCP database with synthetically disrupted ROIs and our results show that OCE-Net is capable of recovering the original FODs by only using information from neighbouring voxels; and demonstrate relevance from both computer vision and neuroscience perspectives.

Acknowledgments. The authors acknowledge the support of an Australian Government Research Training Program (RTP) Scholarship. The authors acknowledge the funding support by the Australia Medical Research Future Fund under Grant (MRF-FAI000085).

Compliance with Ethical Standards. Data used in this work were obtained from the MGH-USC Human Connectome Project (HCP) database.

References

1. Andersson, J.L.R., Sotiropoulos, S.N.: Non-parametric representation and prediction of single-and multi-shell diffusion-weighted MRI data using gaussian processes. Neuroimage **122**, 166–176 (2015)
2. Andersson, J.L.R., Sotiropoulos, S.N.: An integrated approach to correction for off-resonance effects and subject movement in diffusion MR imaging. Neuroimage **125**, 1063–1078 (2016)

3. Breuer, K., Stommel, M., Korte, W.: Analysis and evaluation of fiber orientation reconstruction methods. J. Compos. Sci. **3**(3), 67 (2019)
4. Çiçek, Ö., Abdulkadir, A., Lienkamp, S.S., Brox, T., Ronneberger, O.: 3D U-net: learning dense volumetric segmentation from sparse annotation. In: Ourselin, S., Joskowicz, L., Sabuncu, M.R., Unal, G., Wells, W. (eds.) MICCAI 2016. LNCS, vol. 9901, pp. 424–432. Springer, Cham (2016). https://doi.org/10.1007/978-3-319-46723-8_49
5. Desikan, R.S., et al.: An automated labeling system for subdividing the human cerebral cortex on MRI scans into gyral based regions of interest. Neuroimage **31**(3), 968–980 (2006)
6. Deslauriers-Gauthier, S., et al.: Edema-informed anatomically constrained particle filter tractography. In: Frangi, A.F., Schnabel, J.A., Davatzikos, C., Alberola-López, C., Fichtinger, G. (eds.) MICCAI 2018. LNCS, vol. 11072, pp. 375–382. Springer, Cham (2018). https://doi.org/10.1007/978-3-030-00931-1_43
7. Dhollander, T., et al.: Fixel-based analysis of diffusion MRI: methods, applications, challenges and opportunities. Neuroimage **241**, 118417 (2021)
8. Dhollander, T., Connelly, A.: A novel iterative approach to reap the benefits of multi-tissue CSD from just single-shell (+ b = 0) diffusion MRI data. In: Proceedings of the International Society for Magnetic Resonance in Medicine (ISMRM), vol. 24, p. 3010. Wiley Online Library (2016)
9. Fischl, B.: FreeSurfer. Neuroimage **62**(2), 774–781 (2012)
10. Hu, J., Shen, L., Sun, G.: Squeeze-and-excitation networks. In: Proceedings of the IEEE Conference on Computer Vision and Pattern Recognition (CVPR), pp. 7132–7141 (2018)
11. Jeurissen, B., Tournier, J.D., Dhollander, T., Connelly, A., Sijbers, J.: Multi-tissue constrained spherical deconvolution for improved analysis of multi-shell diffusion MRI data. Neuroimage **103**, 411–426 (2014)
12. Lipp, I., et al.: Tractography in the presence of multiple sclerosis lesions. Neuroimage **209**, 116471 (2020)
13. Ma, Y., et al.: Multiple sclerosis lesion analysis in brain magnetic resonance images: techniques and clinical applications. IEEE J. Biomed. Health Inform. **26**(6), 2680–2692 (2022). https://doi.org/10.1109/JBHI.2022.3151741
14. Schlemper, J., et al.: Attention gated networks: learning to leverage salient regions in medical images. Med. Image Anal. **53**, 197–207 (2019)
15. Smith, R., Skoch, A., Bajada, C.J., Caspers, S., Connelly, A.: Hybrid surface-volume segmentation for improved anatomically-constrained tractography. In: OHBM (2020)
16. Smith, R.E., Tournier, J.D., Calamante, F., Connelly, A.: Anatomically-constrained tractography: improved diffusion MRI streamlines tractography through effective use of anatomical information. Neuroimage **62**(3), 1924–1938 (2012)
17. Smith, R.E., Tournier, J.D., Calamante, F., Connelly, A.: The effects of SIFT on the reproducibility and biological accuracy of the structural connectome. Neuroimage **104**, 253–265 (2015)
18. Smith, R.E., Tournier, J.D., Calamante, F., Connelly, A.: SIFT2: enabling dense quantitative assessment of brain white matter connectivity using streamlines tractography. Neuroimage **119**, 338–351 (2015)
19. Sotiropoulos, S.N., et al.: Effects of image reconstruction on fiber orientation mapping from multichannel diffusion MRI: reducing the noise floor using sense. Magn. Reson. Med. **70**(6), 1682–1689 (2013)

20. Storelli, L., Pagani, E., Preziosa, P., Filippi, M., Rocca, M.A.: Measurement of white matter fiber-bundle cross-section in multiple sclerosis using diffusion-weighted imaging. Mult. Scler. J. **27**(6), 818–826 (2021)
21. Tang, Z., Cabezas, M., Liu, D., Barnett, M., Cai, W., Wang, C.: LG-Net: lesion gate network for multiple sclerosis lesion inpainting. In: de Bruijne, M., et al. (eds.) MICCAI 2021. LNCS, vol. 12907, pp. 660–669. Springer, Cham (2021). https://doi.org/10.1007/978-3-030-87234-2_62
22. Tournier, J.D., Calamante, F., Connelly, A.: MRtrix: diffusion tractography in crossing fiber regions. Int. J. Imaging Syst. Technol. **22**(1), 53–66 (2012)
23. Tournier, J.D., et al.: MRtrix3: a fast, flexible and open software framework for medical image processing and visualisation. Neuroimage **202**, 116137 (2019)
24. Tournier, J.D., Calamante, F., Connelly, A.: Robust determination of the fibre orientation distribution in diffusion MRI: non-negativity constrained super-resolved spherical deconvolution. Neuroimage **35**(4), 1459–1472 (2007)
25. Tournier, J.D., Calamante, F., Connelly, A., et al.: Improved probabilistic streamlines tractography by 2nd order integration over fibre orientation distributions. In: Proceedings of the International Society for Magnetic Resonance in Medicine, vol. 1670. Wiley Online Library (2010)
26. Tournier, J.D., Calamante, F., Gadian, D.G., Connelly, A.: Direct estimation of the fiber orientation density function from diffusion-weighted MRI data using spherical deconvolution. Neuroimage **23**(3), 1176–1185 (2004)
27. Tustison, N.J., et al.: N4ITK: improved N3 bias correction. IEEE Trans. Med. Imaging **29**(6), 1310–1320 (2010)
28. Van Essen, D.C., Smith, S.M., Barch, D.M., Behrens, T.E.J., Yacoub, E., Ugurbil, K.: The WU-Minn human connectome project: an overview. Neuroimage **80**, 62–79 (2013)
29. Yeh, C.H., Smith, R.E., Dhollander, T., Calamante, F., Connelly, A.: Connectomes from streamlines tractography: assigning streamlines to brain parcellations is not trivial but highly consequential. Neuroimage **199**, 160–171 (2019)
30. Yu, J., Lin, Z., Yang, J., Shen, X., Lu, X., Huang, T.S.: Free-form image inpainting with gated convolution. In: Proceedings of the IEEE International Conference on Computer Vision (ICCV), pp. 4471–4480 (2019)
31. Zeng, R., et al.: FOD-net: a deep learning method for fiber orientation distribution angular super resolution. Med. Image Anal. **79**, 102431 (2022)

Fitting a Directional Microstructure Model to Diffusion-Relaxation MRI Data with Self-supervised Machine Learning

Jason P. Lim[1], Stefano B. Blumberg[1,2], Neil Narayan[1], Sean C. Epstein[1], Daniel C. Alexander[1], Marco Palombo[1,3,4], and Paddy J. Slator[1](✉)

[1] Centre for Medical Image Computing, Department of Computer Science, University College London, London, UK
p.slator@ucl.ac.uk

[2] Centre for Artificial Intelligence, Department of Computer Science, University College London, London, UK

[3] Cardiff University Brain Research Imaging Centre (CUBRIC), School of Psychology, Cardiff University, Cardiff, UK

[4] School of Computer Science and Informatics, Cardiff University, Cardiff, UK

Abstract. Machine learning is a powerful approach for fitting microstructural models to diffusion MRI data. Early machine learning microstructure imaging implementations trained regressors to estimate model parameters in a supervised way, using synthetic training data with known ground truth. However, a drawback of this approach is that the choice of training data impacts fitted parameter values. Self-supervised learning is emerging as an attractive alternative to supervised learning in this context. Thus far, both supervised and self-supervised learning have typically been applied to isotropic models, such as intravoxel incoherent motion (IVIM), as opposed to models where the directionality of anisotropic structures is also estimated. In this paper, we demonstrate self-supervised machine learning model fitting for a directional microstructural model. In particular, we fit a combined T1-ball-stick model to the multidimensional diffusion (MUDI) challenge diffusion-relaxation dataset. Our self-supervised approach shows clear improvements in parameter estimation and computational time, for both simulated and in-vivo brain data, compared to standard non-linear least squares fitting. Code for the artificial neural net constructed for this study is available for public use from the following GitHub repository: https://github.com/jplte/deep-T1-ball-stick.

Keywords: Microstructure imaging · Machine learning · Self-supervised learning

1 Introduction

Microstructure imaging aims to quantify features of the tissue microstructure from in-vivo MRI [1]. Historically, microstructure imaging utilised diffusion MRI

S. Cetin-Karayumak et al. (Eds.): CDMRI 2022, LNCS 13722, pp. 77–88, 2022.
https://doi.org/10.1007/978-3-031-21206-2_7

(dMRI) data. Recently, combined diffusion-relaxation MRI - where relaxation-encoding parameters such as inversion time (TI) and echo time (TE) are varied alongside diffusion-encoding parameters such as b-value and gradient direction - has been emerging as an extension [23]. The typical approach to estimating tissue microstructure from such diffusion or diffusion-relaxation data is multi-compartment modelling, which utilises signal models comprising linear combinations of multiple compartments - such as balls, sticks, zeppelins, and spheres - each representing a distinct tissue geometry [21].

Multi-compartment microstructure models are usually fit to the data with non-linear least squares (NLLS) algorithms. However, these can be computationally expensive and are prone to local minima, necessitating grid searches or parameter cascading [9] to seek global minima. Machine learning is a powerful alternative. Thus far, most machine learning microstructure model fitting approaches have used supervised learning [7,10,14,17–20]. However, a crucial limitation is that the distribution of training data significantly affects fitted parameters [8,12]. It has also proved difficult to estimate directional parameters, such as fibre direction, with existing machine learning methods instead directly estimating rotationally invariant parameters, such as mean diffusivity, fractional anisotropy, mean kurtosis, and orientation dispersion. This may be due to the difficulty of constructing a training dataset that adequately samples the high-dimensional parameter space, and/or complications due to the periodicity of angular parameters.

Self-supervised (sometimes imprecisely called unsupervised in the microstructure imaging context) learning is an alternative with the potential to address these limitations. Self-supervised algorithms learn feature representations from the input data by inferring supervisory constraints from the data itself. For microstructure imaging, self-supervised learning has been implemented with voxelwise fully connected artificial neural networks (ANNs). However, thus far self-supervised microstructure imaging has been limited to isotropic models [6,11], including many intravoxel incoherent motion (IVIM) MRI examples [2,8,13,25,26]. To our knowledge, self-supervised model fitting has not yet been demonstrated for directional microstructural models.

In this paper, we fit an extended T1-ball-stick model to diffusion-relaxation MRI data using self-supervised machine learning and demonstrate several advantages of this approach over classical NLLS, such as higher precision and faster computational time.

2 Methods

2.1 Microstructure Model

As this is a first attempt at fitting directional multi-compartment models with self-supervised learning, we choose a simple model - the ball-stick model first proposed by Behrens et al. [3]. According to the ball-stick model, the expression for the normalized signal decay is

$$S(b, \mathbf{g}) = f \exp\left(-b\lambda_{||}(\mathbf{g}.\mathbf{n})\right) + (1-f)\exp\left(-b\lambda_{iso}\right) \tag{1}$$

where b is the b-value, $\mathbf{g}$ is the gradient direction, $\lambda_{||}$ and λ_{iso} are the parallel and isotropic diffusivities of the stick and ball respectively, and $\mathbf{n}$ is the stick orientation, which we parameterise using polar coordinates. The relationship between Cartesian and polar coordinates is $n = [\sin\theta\cos\phi, \sin\theta\sin\phi, \cos\theta]$ where $\phi \in [0, \pi]$ and $\theta \in [-\pi, \pi]$.

We extend the ball-stick model to account for T1 relaxation time, by assuming the ball and stick compartments have separate T1 times, represented by $T1_{ball}$ and $T1_{stick}$ respectively. Note that we assume a single T2 for both compartments, so the volume fraction f will be affected by the T2 of each compartment. Given a combined T1 inversion recovery [5] and diffusion MRI experiment, where inversion time (TI), b-value and gradient direction are simultaneously varied, we can fit the following T1-ball-stick equation

$$S(b, g, T_I, T_R) = f \exp\left(-b\lambda_{||}(\mathbf{g}.\mathbf{n})\right) \left|1 - 2\exp\left(-\frac{T_I}{T1_{stick}}\right)\exp\left(-\frac{T_R}{T1_{stick}}\right)\right| + (1-f)\exp\left(-b\lambda_{iso}\right)\left|1 - 2\exp\left(-\frac{T_I}{T1_{ball}}\right)\exp\left(-\frac{T_R}{T1_{ball}}\right)\right| \quad (2)$$

In this work, we first fit this model to combined T1-diffusion data with standard NLLS, then demonstrate self-supervised fitting with an ANN. We first describe the data, then the model fitting techniques.

2.2 Combined T1-Diffusion in Vivo Data

We utilise in-vivo data from 5 healthy volunteers (3 F, 2 M, age = 19–46 years), acquired from the 2019 multidimensional diffusion (MUDI) challenge [22]. The acquisition sequence comprises simultaneous diffusion, inversion recovery (giving T1 contrast), and multi-echo gradient echo (giving T2* contrast) measurements. We chose to ignore the subsection of the data that is sensitive to T2* by only included signals captured with the lowest echo time (80 ms). This is since the two higher TEs have very low signal intensity and the 3 TEs have a small range, and thus there is limited T2* information in the data. Our subsequent description hence only refers to the subsection of the data with TE = 80 ms.

The datasets were obtained using a clinical 3T Philips Achieva scanner (Best, Netherlands) with a 32-channel adult headcoil. Each scan includes 416 volumes distributed over five b-shells, b $\in \{0, 500, 1000, 2000, 3000\}$ s/mm^2, with 16 uniformly spread directions, and 28 inversion times (TI) $\in$ [20, 7322] ms. For all datasets, the following parameters were fixed: repetition time TR = 7.5 s, resolution = 2.5 mm isotropic, FOV = 220×230×140 mm, SENSE = 1.9, halfscan = 0.7, multiband factor 2, total acquisition time 52 min (including preparation time).

The MUDI data has already undergone standard pre-processing, see [22] for full details. Upon inspection, we noted that the lowest (20 ms) and highest (7322 ms) inversion times, which comprise 7.14% of the data, were clearly dominated by noise and/or artifacts (see Fig. 1). We therefore removed them from the data prior to model fitting, leaving 416 MRI volumes in total. After removing these TIs, the dataset contains 26 TIs $\in\in$ [176, 4673] ms.

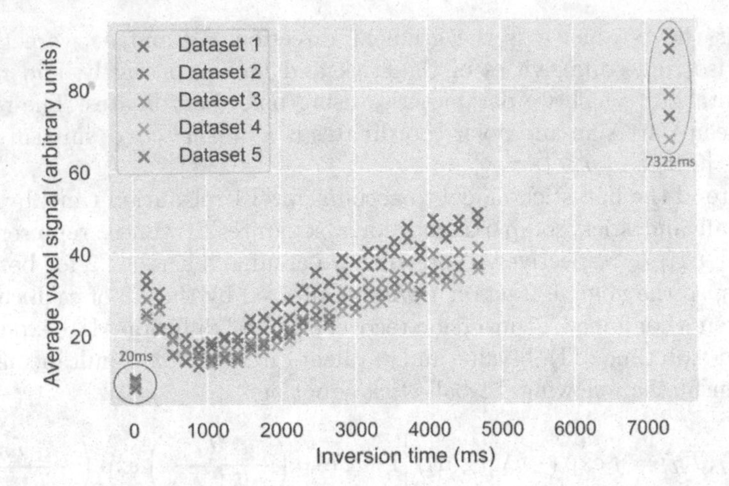

Fig. 1. Average signal of all b = 0 voxels within the brain mask at each inversion time. Note the outlying signals at the smallest (20 ms, blue) and largest (7322 ms, green) inversion times. We removed these from the data before fitting. (Color figure online)

We normalised each voxel's data independently, by dividing by the signal generated from the b = 0 volume with the highest TI, i.e. the volume with the highest expected signal. We then removed all background voxels using the brain mask provided with the MUDI data.

2.3 Simulated Data

In-vivo MRI data does not have ground truth tissue-related parameters values, making it hard to quantitatively assess the accuracy of model fitting. We thus simulated 100,000 synthetic signals using the T1-ball-stick model signal equation (Eq. (2)). We used the same 416 acquisition parameters (b-values, gradient directions and inversion times) as our reduced MUDI dataset. The ground truth values of $\lambda_{||}$, λ_{iso}, $T1_{ball}$ and $T1_{stick}$ were sampled randomly from physically-plausible ranges (Table 1). Note that we choose units so that parameter values are close to 1; this prevents having to normalise parameter values before training neural networks. Complex Gaussian noise was added to simulate the Rician distribution of noisy MRI data [16].

2.4 Non-linear Least Squares Fitting

The modified T1-ball-stick model (Eq. (2)) was fit with non-linear least squares by modifying the open source diffusion microstructure imaging in python (dmipy) toolbox [9], with parameter constraints as in Table 1. Specifically, we used the "brute2fine" function, which uses a brute force grid search followed by non-linear optimisation.

Table 1. Constraints on T1-ball-stick parameters for simulating data and model fitting.

Parameter	Minimum	Maximum
λ_{iso}	0.1 $\mu m^2/ms$	3.0 $\mu m^2/ms$
$\lambda_{\|\|}$	0.1 $\mu m^2/ms$	3.0 $\mu m^2/ms$
f	0	1
$T1_{ball}$	0.01 s	5 s
$T1_{stick}$	0.01 s	5 s
θ	0	π
ϕ	$-\pi$	π

2.5 Self-supervised Model Fitting

We built an ANN to generate estimates for the T1-ball-stick parameters. The network comprises an input layer, 3 fully connected hidden layers and an output layer, (see Fig. 2). The input layer and hidden layers each have 416 nodes - mirroring the 416 MRI volumes. The final layer has 7 output neurons, one for each parameter of interest. The normalised signal from a single voxel of the MRI data, S, which comprises 416 measurements, is fed into the input layer and passed through the ANN. The output layer is fed forward into the T1-ball-stick model equation, giving a synthetic signal $\hat{S}$. Training loss is the mean squared error between input (S) and synthetic ($\hat{S}$) MRI signals across all voxels passed through the ANN.

We implemented the ANN on Python 3.9.5 using PyTorch [21] 1.10.0. For both simulated and in-vivo data, we used the Adam optimiser [15] with learning rate 0.0001, batch size 128, and dropout [24] with rate 0.5. Parameter constraints (Table 1) were imposed using PyTorch's clamp feature, which converts any value outside the bounds to the value closest to it within the boundary. Following [2], we trained the network with patience 10, i.e. until there 10 consecutive epochs without loss improvement. As the self-supervised approach estimates T1-ball-stick parameters directly from the data, we didn't use a train-test split. Instead, the ANN was trained on each dataset separately.

3 Results

3.1 Simulated Data

The T1-ball-stick model was successfully fit to the simulated signals using both the ANN and NLLS. Scatter plots and Pearson correlation coefficients of parameter estimates against ground truth values are shown in Fig. 3 (note that we don't report correlations for θ and ϕ as the values are confounded by the periodicity of these angular parameters). Correlation coefficients for the ANN fits are higher for all model parameters, with coefficients above 0.9 for all parameters except $\lambda_{||}$. NLLS correlation coefficients for T1 relaxation time are particularly low.

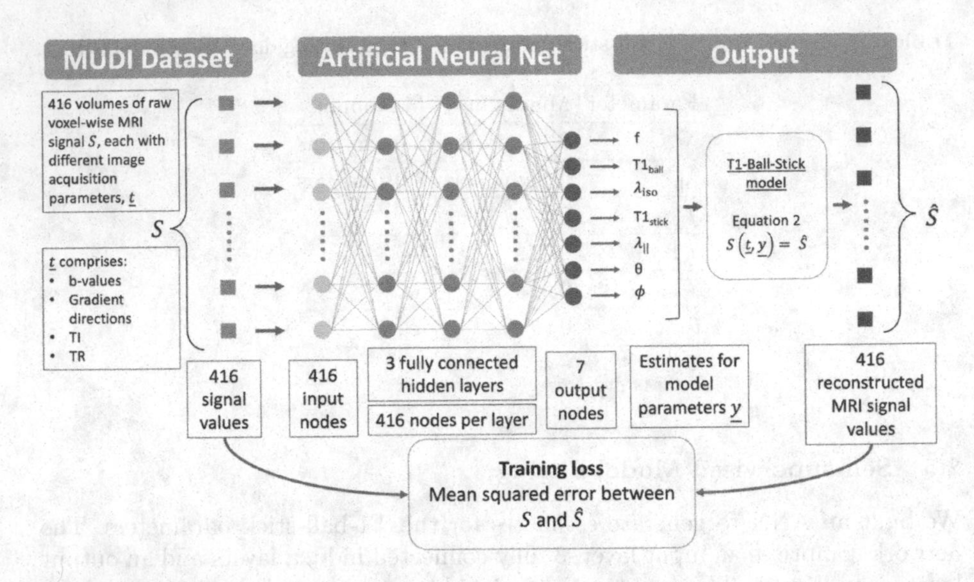

Fig. 2. Our ANN for T1-ball-stick fitting and the flow of data through it.

3.2 Real Data

Figures 4 and 5 show T1-ball-stick model fits for all MUDI subjects. ANN parameter maps are qualitatively less noisy and show more anatomically plausible contrast than NLLS. This is particularly clear for $T1_{ball}$, $\lambda_{||}$ and direction encoded colour (DEC) maps. All 5 MUDI subjects showed similar trends. ANN inferred higher and lower values than NLLS for $T1_{ball}$ and $T1_{stick}$ respectively. The ANN clearly shows highest $\lambda_{||}$ values in the corpus callosum, while the Dmipy fit has high $\lambda_{||}$ values in many places. λ_{iso} maps are generally similar across both methods.

Average time taken for ANN fits on real data was 1966s, compared to 8833 s in Dmipy, meaning ANN was 77.25% faster than NLLS on average. All model fits were performed on a 2017 Macbook Pro's central processing unit (3.1 GHz Dual-Core Intel I5-7267U).

4 Discussion

This study demonstrates self-supervised microstructure imaging for a combined T1-ball-stick model. Our ANN approach is faster and more precise that conventional NLLS model fitting. In the ANN model fits, parametric maps show plausible estimates for both diffusivity ($\lambda_{||}$, λ_{iso}) and relaxation ($T1_{ball}$ and $T1_{stick}$), whereas some NLLS maps, particularly $T1_{ball}$, show dubious contrast.

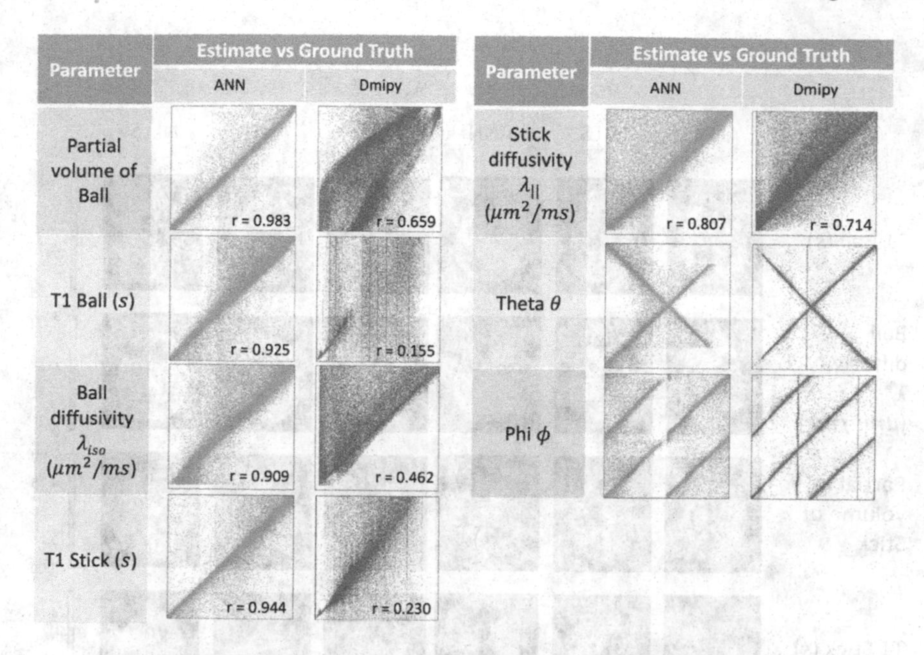

Fig. 3. NLLS and ANN T1-ball-stick estimates against ground truth values, after fitting the model to simulated signals. r denotes the Pearson correlation coefficient.

The ANN model fits potentially reveal more accurate tissue information than NLLS counterparts. Throughout the brain, white matter T1 times are expected to be around 0.7–0.9 s [4]. The ANN estimates fall within this range - the $T1_{stick}$ voxels displayed in Figs. 4 and 5 average to 0.87 s - while NLLS estimates are higher, with some regions reaching 4 s. T1 times in the CSF are expected to be around 4 s [4], which is reflected in the ANN $T1_{ball}$ estimate, but not in the NLLS estimate, where it is approximately 0 s (Figs. 4 and 5).

These observations match those in the simulations (Fig. 3), with NLLS correlation being extremely poor in both T1 times. In line with our qualitative analysis of MUDI data fits, Figure 3 shows that the ANN outperforms Dmipy in every parameter, except potentially for θ and ϕ, whose correlations are not straightforward to quantify due to their periodicity. Direction encoded colour (DEC) maps are less noisy for ANN fits, with less visible noise, and are hence potentially more useful for tractography.

The ANN approach was faster in all datasets, with a 77% improvement in time on average. This will hopefully improve the feasibility of utilising similar modelling in clinical scenarios. However, the ANN would still require retraining for every new dataset. A possible next step would be to explore the viability of using multiple datasets to train an ANN to be generalisable to unseen data, then fine-tuning the network on each new dataset, effectively combining advantages of supervised and self-supervised learning, as recently demonstrated by Epstein et al. [8].

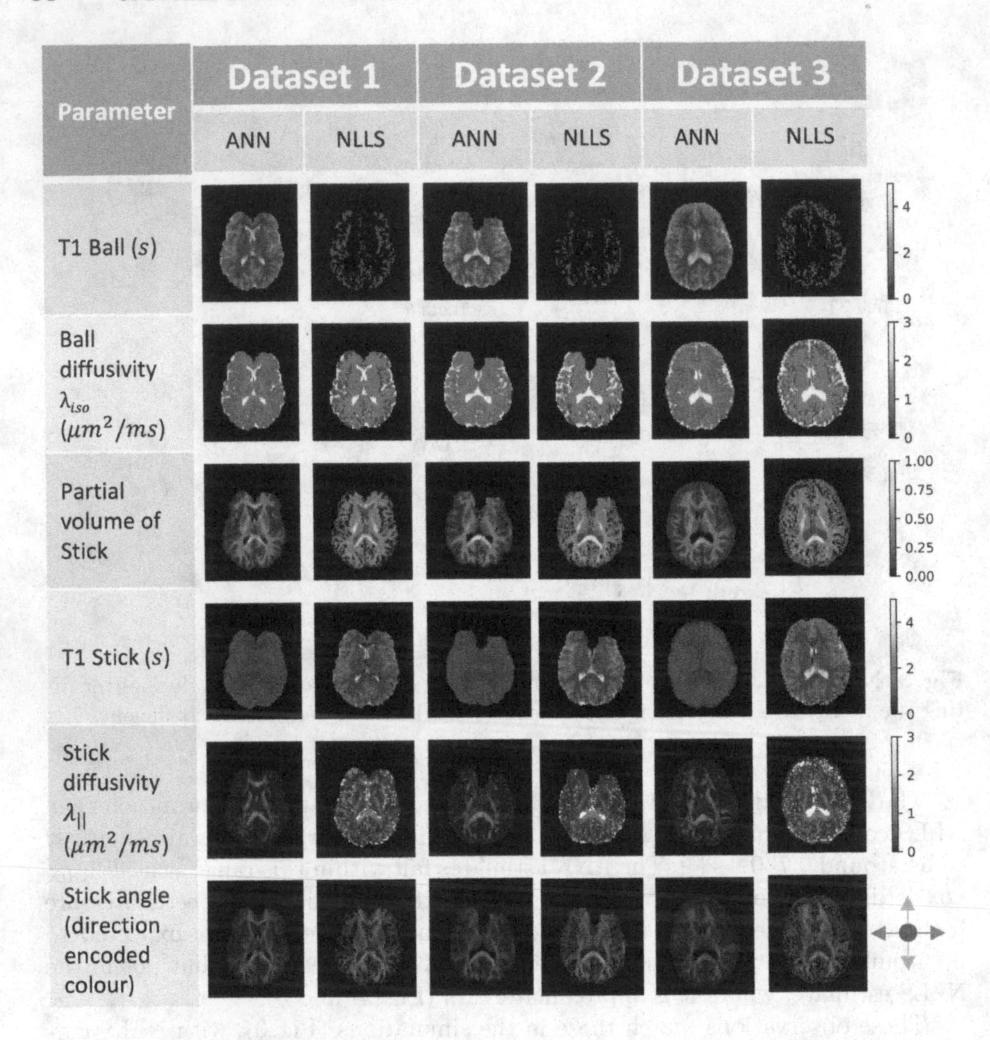

Fig. 4. Parameter maps for all models fit to the first 3 MUDI datasets using both Dmipy and ANN methods. Each parameter map is a cross-sectional view of the brain. The maps presented are generated from the middle Z-axis slice.

If successful, this could significantly reduce time taken to generate parameter estimates for new patients, but possibly at the cost of accuracy.

Whilst the ANN outperforms NLLS in our experiments, we applied NLLS "out of the box" without focusing on improving the fitting. The NLLS fits could be improved - e.g. the vertical lines in Fig. 3 are likely local minima. . In future, we could use a larger grid in the grid search stage, although this can quickly lead to infeasible computational times, or initialise the NLLS fit from "reasonable" parameter values. Whilst these would likely improve the NLLS fits, the fact that our self-supervised ANN bypasses these ad-hoc tuning steps presents a significant advantage. We also only compare parameter estimates with the ground truth

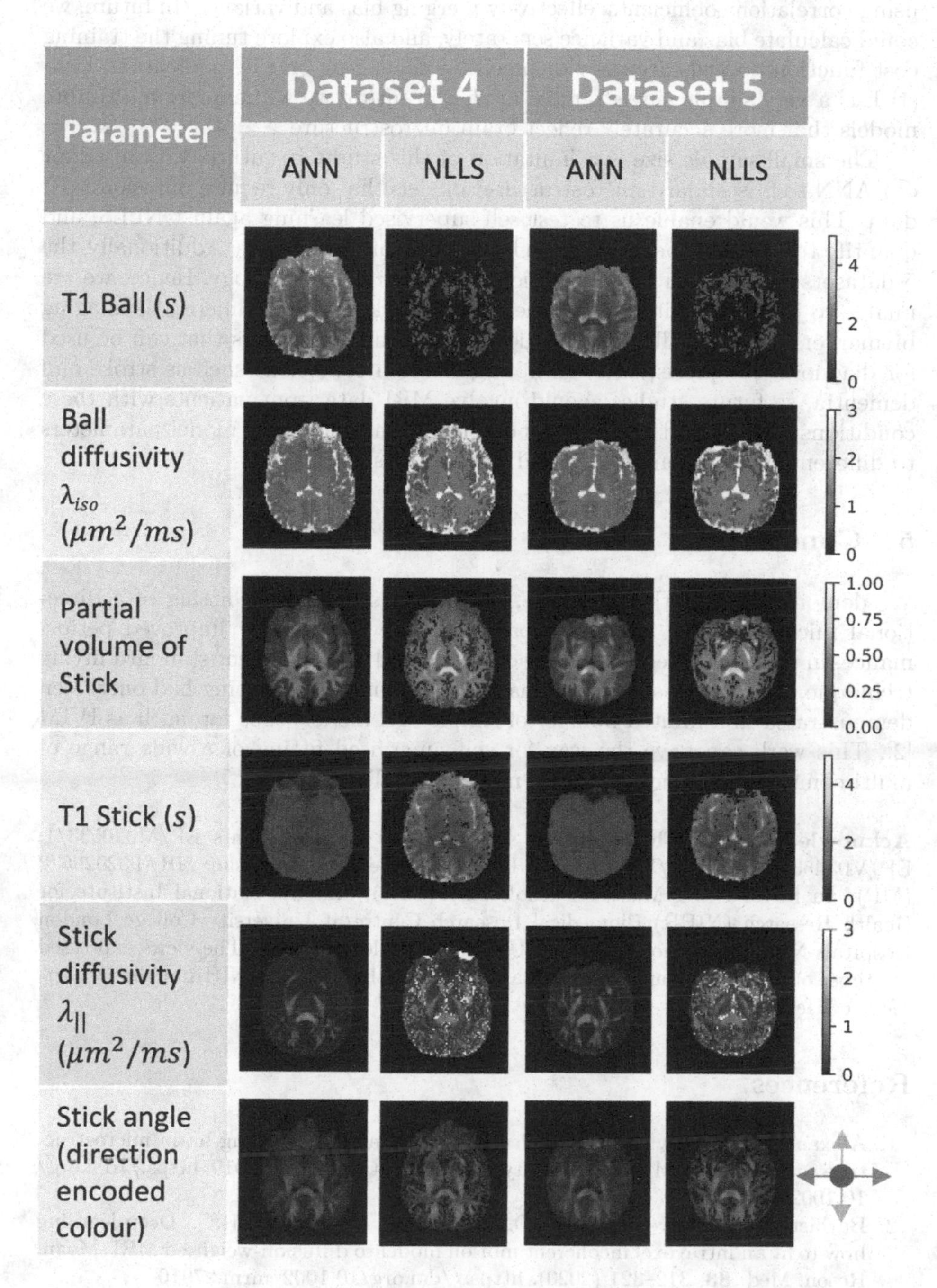

Fig. 5. As Fig. 4 but for the remaining 2 MUDI datasets.

using correlation coefficients, effectively merging bias and variance. In future, we could calculate bias and variance separately, and also explore tuning the training cost function towards accuracy or precision depending on the application. Ball-stick is a very simple single bundle model, in future we will explore multi-fibre models that more accurately reflect brain microstructure.

The small sample size is a limitation of this study. In future, we can adapt our ANN to fit standard microstructural models that only require diffusion MRI data. This would enable us to test self-supervised learning against NLLS, and quantify test-retest repeatability, on large open source datasets. Additionally, the 5 datasets are all from healthy patients with normal physiology. Hence, we are unable to judge the suitability of the T1-ball-stick model parameters as imaging biomarkers. The overall goal is to identify imaging biomarkers that can be used for diagnosis, prognosis, and monitoring of brain conditions such as stroke and dementia, so future studies should involve MRI data from patients with these conditions. This would also help us determine thresholds for model parameters to differentiate between healthy and diseased tissue.

5 Conclusion

We demonstrate, for the first time, self-supervised learning fitting of a directional microstructural model, T1-ball-stick. We show vastly improved performance, in terms of speed and accuracy, compared to the current standard fitting technique, NLLS. Self-supervised machine learning model fitting had only been demonstrated in a limited number of simple MC models thus far, such as IVIM [2]. This work can pave the way for self-supervised fitting of a wide range of multi-compartment microstructure models to MRI data.

Acknowledgments. This work was supported by EPSRC grants EP/M020533/1, EP/V034537/1, and EP/R014019/1; UKRI Future Leaders Fellowship MR/T020296/2 (MP); an EPRSC and Microsoft scholarship (SBB); and the National Institute for Health Research (NIHR) Biomedical Research Centre at University College London Hospitals NHS Foundation Trust and University College London. The views expressed are those of the authors and not necessarily those of the NHS, the NIHR or the Department of Health.

References

1. Alexander, D.C., Dyrby, T.B., Nilsson, M., Zhang, H.: Imaging brain microstructure with diffusion MRI: practicality and applications, April 2019. https://doi.org/10.1002/nbm.3841
2. Barbieri, S., Gurney-Champion, O.J., Klaassen, R., Thoeny, H.C.: Deep learning how to fit an intravoxel incoherent motion model to diffusion-weighted MRI. Magn. Reson. Med. **83**, 312–321 (2020). https://doi.org/10.1002/mrm.27910
3. Behrens, T.E., et al.: Characterization and propagation of uncertainty in diffusion-weighted MR imaging. Magn. Reson. Med. **50**, 1077–1088 (2003). https://doi.org/10.1002/mrm.10609

4. Bojorquez, J.Z., Bricq, S., Acquitter, C., Brunotte, F., Walker, P.M., Lalande, A.: What are normal relaxation times of tissues at 3t?, January 2017. https://doi.org/10.1016/j.mri.2016.08.021
5. Bydder, G.M., Young, I.R.: MR imaging: clinical use of the inversion recovery sequence. J. Comput. Assist. Tomogr. **9**, 659–675 (1985)
6. de Almeida Martins, J.P., et al.: Neural networks for parameter estimation in microstructural MRI: application to a diffusion-relaxation model of white matter. NeuroImage **244**, 118601 (2021). https://doi.org/10.1016/j.neuroimage.2021.118601
7. Diao, Y., Jelescu, I.O.: Parameter estimation for WMTI-Watson model of white matter using encoder- decoder recurrent neural network. ArXiv Preprint (2022)
8. Epstein, S.C., Bray, T.J.P., Hall-Craggs, M., Zhang, H.: Choice of training label matters: how to best use deep learning for quantitative MRI parameter estimation. ArXiv Preprint, May 2022. https://doi.org/10.48550/arxiv.2205.05587
9. Fick, R.H.J., Wassermann, D., Deriche, R.: The Dmipy toolbox: diffusion MRI multi-compartment modeling and microstructure recovery made easy. Front. Neuroinform. **13**, 64 (2019). https://doi.org/10.3389/fninf.2019.00064
10. Golkov, V., et al.: q-space deep learning: twelve-fold shorter and model-free diffusion MRI scans. IEEE Trans. Med. Imaging **35**(5), 1344–1351 (2016). https://doi.org/10.1109/TMI.2016.2551324
11. Grussu, F., Battiston, M., Palombo, M., Schneider, T., Wheeler-Kingshott, C.A., Alexander, D.C.: Deep learning model fitting for diffusion-relaxometry: a comparative study. In: Computational Diffusion MRI (CDMRI) 2021, pp. 159–172. Springer (2021). https://doi.org/10.1007/978-3-030-73018-5_13
12. Gyori, N.G., Palombo, M., Clark, C.A., Zhang, H., Alexander, D.C.: Training data distribution significantly impacts the estimation of tissue microstructure with machine learning. Magn. Reson. Med. **87**, 932–947 (2022). https://doi.org/10.1002/mrm.29014
13. Kaandorp, M.P., et al.: Improved unsupervised physics-informed deep learning for intravoxel incoherent motion modeling and evaluation in pancreatic cancer patients. Magn. Reson. Med. **86**(4), 2250–2265 (2021). https://doi.org/10.1002/mrm.28852
14. Kerkelä, L., Seunarine, K., Henriques, R.N., Clayden, J.D., Clark, C.A.: Improved reproducibility of diffusion kurtosis imaging using regularized non-linear optimization informed by artificial neural networks (2022). arxiv:2203.07327
15. Kingma, D.P., Ba, J.: Adam: A method for stochastic optimization (2014). arxiv:1412.6980
16. Koay, C.G., Basser, P.J.: Analytically exact correction scheme for signal extraction from noisy magnitude MR signals. J. Magn. Reson. **179**, 317–322 (2006). https://doi.org/10.1016/j.jmr.2006.01.016
17. Li, Z., et al.: Fast and robust diffusion kurtosis parametric mapping using a three-dimensional convolutional neural network. IEEE Access **7**, 71398–71411 (2019). https://doi.org/10.1109/ACCESS.2019.2919241
18. Nedjati-Gilani, G.L., et al.: Machine learning based compartment models with permeability for white matter microstructure imaging. NeuroImage **150**, 119–135 (2017). https://doi.org/10.1016/j.neuroimage.2017.02.013
19. Palombo, M., et al.: SANDI: a compartment-based model for non-invasive apparent soma and neurite imaging by diffusion MRI. NeuroImage **215**, 116835 (2020). https://doi.org/10.1016/j.neuroimage.2020.116835

20. Palombo, M., et al.: Joint estimation of relaxation and diffusion tissue parameters for prostate cancer grading with relaxation-VERDICT MRI. medRxiv 1(165) (2021). https://doi.org/10.1101/2021.06.24.21259440
21. Panagiotaki, E., Schneider, T., Siow, B., Hall, M.G., Lythgoe, M.F., Alexander, D.C.: Compartment models of the diffusion MR signal in brain white matter: A taxonomy and comparison. Neuroimage **59**(3), 2241–2254 (2012). https://doi.org/10.1016/j.neuroimage.2011.09.081
22. Pizzolato, M., et al.: Acquiring and predicting multidimensional diffusion (MUDI) data: an open challenge (2020). https://doi.org/10.1007/978-3-030-52893-5_17
23. Slator, P.J., et al.: Combined diffusion-relaxometry microstructure imaging: current status and future prospects, December 2021. https://doi.org/10.1002/mrm.28963
24. Srivastava, N., Hinton, G., Krizhevsky, A., Salakhutdinov, R.: Dropout: a simple way to prevent neural networks from overfitting (2014)
25. Vasylechko, S.D., Warfield, S.K., Afacan, O., Kurugol, S.: Self-supervised IVIM DWI parameter estimation with a physics based forward model. Magn. Reson. Med. **00**, 1–11 (2021). https://doi.org/10.1002/mrm.28989
26. Zhou, X.X., et al.: An unsupervised deep learning approach for dynamic-exponential intravoxel incoherent motion MRI modeling and parameter estimation in the liver. J. Magn. Reson. Imaging (2022). https://doi.org/10.1002/jmri.28074

Stepwise Stochastic Dictionary Adaptation Improves Microstructure Reconstruction with Orientation Distribution Function Fingerprinting

Patryk Filipiak[1(✉)], Timothy Shepherd[1], Lee Basler[2], Anthony Zuccolotto[2], Dimitris G. Placantonakis[3], Walter Schneider[4], Fernando E. Boada[5], and Steven H. Baete[1]

[1] Center for Advanced Imaging Innovation and Research (CAI2R), Department of Radiology, NYU Langone Health, New York, NY, USA
patryk.filipiak@nyulangone.org
[2] Psychology Software Tools, Inc., Pittsburgh, PA, USA
[3] Department of Neurosurgery, Perlmutter Cancer Center, Neuroscience Institute, Kimmel Center for Stem Cell Biology, NYU Langone Health, New York, NY, USA
[4] University of Pittsburgh, Pittsburgh, PA, USA
[5] Radiological Sciences Laboratory and Molecular Imaging Program at Stanford, Department of Radiology, Stanford University, Stanford, CA, USA

Abstract. Fitting of the multicompartment biophysical model of white matter is an ill-posed optimization problem. One approach to make it computationally tractable is through Orientation Distribution Function (ODF) Fingerprinting. However, the accuracy of this method relies solely on ODF dictionary generation mechanisms which either sample the microstructure parameters on a multidimensional grid or draw them randomly with a uniform distribution. In this paper, we propose a stepwise stochastic adaptation mechanism to generate ODF dictionaries tailored specifically to the diffusion-weighted images in hand. The results we obtained on a diffusion phantom and in vivo human brain images show that our reconstructed diffusivities are less noisy and the separation of a free water fraction is more pronounced than for the prior (uniform) distribution of ODF dictionaries.

Keywords: Brain microstructure · White matter · ODF Fingerprinting · Diffusion MRI · Stochastic optimization

1 Introduction

Brain White Matter (WM) microstructure features are reconstructed in vivo from Diffusion Weighted Images (DWIs) by fitting biophysical models [4,9,17] of acquired signal. In a typical scenario, this boils down to solving a non-convex optimization problem with multiple local optima [10] which is computationally challenging.

S. Cetin-Karayumak et al. (Eds.): CDMRI 2022, LNCS 13722, pp. 89–100, 2022.
https://doi.org/10.1007/978-3-031-21206-2_8

One numerical approach to this problem uses Orientation Distribution Function Fingerprinting (ODF-FP) [2] to find near-optimal solutions in linear time by matching ODFs of the acquired signal with the elements of a precomputed ODF dictionary. However, the accuracy of this approach relies solely on the ODF dictionary generation mechanism which either samples the microstructure parameters on a multidimensional grid [2] or draws them randomly with a uniform distribution [6]. Both these techniques lack specificity due to the inherent assumption that every element of an ODF dictionary is equally likely to be found in the dataset.

In this paper, we propose a stepwise stochastic adaptation mechanism to generate ODF dictionaries tailored specifically to the DWIs in hand. Our approach implements an Estimation of Distribution Algorithm (EDA) [7,14] to statistically infer posterior distribution of ODF dictionary elements. By gradually improving the prior uniform distribution of microstructure parameters, our algorithm adapts the sampling mechanism of the ODF dictionary to the acquired DWIs in an unsupervised, data-driven manner. Through this, we address the lack of specificity in the original ODF dictionary design [6], which in practice led to storing multiple ODF fingerprints that were unlikely to be selected.

We present the results obtained on a diffusion phantom and in vivo human brain images showing that our approach improves microstructure parameters estimation with ODF-FP. Our reconstructed diffusivities are less noisy and the separation of a free water fraction is more pronounced. This leads to more accurate approximation of clinically significant microstructure features attributed to axonal loss [4], inflammation [20], or demyelination [11].

2 Methods

In this study, we reconstructed WM microstructure parameters using ODF-FP. Note that our method did not impose any particular definition of ODF. For brevity, though, we considered the so-called *diffusion ODF* variant [23]. From now on, we will refer to it simply as ODF.

2.1 Biophysical Diffusion Model

We used the multicompartment diffusion model [8] defined as

$$S(b) = S(0) \cdot \left[p_{iso} e^{-bD_{iso}} + \sum_{i=1}^{N} p^{(i)} \kappa^{(i)}(b, \mathbf{g} \cdot \mathbf{n}^{(i)}) \right], \tag{1}$$

where $S(0)$ is the signal without diffusion encoding ($b = 0$), while the contribution of i-th fiber ($i = 1, \ldots, N$) is

$$\begin{aligned} \kappa^{(i)}(b, \mathbf{g} \cdot \mathbf{n}^{(i)}) = &\, f^{(i)} e^{-bD_{a,\parallel}^{(i)} (\mathbf{g}\cdot\mathbf{n}^{(i)})^2} \\ &+ \left(1 - f^{(i)}\right) e^{-bD_{e,\parallel}^{(i)} (\mathbf{g}\cdot\mathbf{n}^{(i)})^2 - bD_{e,\perp}^{(i)} \left(1-(\mathbf{g}\cdot\mathbf{n}^{(i)})^2\right)}, \end{aligned} \tag{2}$$

where $\mathbf{n}^{(i)} \in \mathbb{R}^3$ is the fiber orientation and $\mathbf{g} \in \mathbb{R}^3$ is the direction of the diffusion encoding gradient. The compartment volumes of free water $p_{iso} \in [0, 0.8]$ and neurites $p^{(i)} \geq 0.1$ sum up to 1. The fraction sizes are $f^{(i)} \in [0, 0.8]$. The ranges of diffusivities are as follows: free water $D_{iso} \in [2, 3]$, intra-axonal $D_{a,\|}^{(i)} \in [1.5, 2.5]$, and extra-axonal $D_{e,\|}^{(i)} \in [1.5, 2.5]$, $D_{e,\perp}^{(i)} \in [0.5, 1.5] \cdot 10^{-9}$ m^2/s, assuming that $D_{a,\|}^{(i)} \geq D_{e,\|}^{(i)}$ as advocated in [9].

2.2 Orientation Distribution Function Fingerprinting

Throughout this study, we maintained the following two types of ODF dictionaries designed for our two datasets:

(a) **phantom dataset**—a simplified dictionary of 10^4 elements, each of them limited to $N \leq 2$ fibers per voxel and equal fiber fractions, i.e. $D_{a,\|} = D_{a,\|}^{(i)}$, $D_{e,\|} = D_{e,\|}^{(i)}$, $D_{e,\perp} = D_{e,\perp}^{(i)}$, and $f = f^{(i)}$ for $i = 1, 2$.

(b) **in vivo dataset**—a dictionary of 10^6 elements, each of them limited to $N \leq 3$ fibers per voxel as suggested by Jeurissen et al. [12], without any simplifications of the diffusion model.

In either case, the b-values and the diffusion sampling directions $\mathbf{g}$ matched the data acquisition protocols defined later in Subsect. 2.4.

For matching of ODF fingerprints, we used a k-point tessellation of a unit hemisphere (with $k = 321$) to discretize ODFs. Having this, we applied the matching formula [2,6] defined as

$$\widetilde{\mathbf{x}} = \arg\max_{\mathbf{d} \in \mathcal{D}} \left(\log \mathbf{x}^T \mathbf{d} - N \cdot \lambda \right), \tag{3}$$

where $\mathbf{x} \in \mathbb{R}^k$ is the fingerprint of a given ODF computed from the acquired signal, $\widetilde{\mathbf{x}}$ is its best-fitting representative among the elements $\mathbf{d}$ of the ODF dictionary $\mathcal{D}$, and $\lambda > 0$ is the penalty factor to limit the number of ODF peaks. Note that the formula in Eq. 3 is a modification of the Akaike information criterion [1], where $\mathbf{x}^T \mathbf{d}$ approximates the likelihood function of the diffusion model, while the $N \cdot \lambda$ component discourages overfitting. Here, we chose the empirical values of $\lambda = 2 \cdot 10^{-4}$ in our phantom study and $\lambda = 1 \cdot 10^{-3}$ in vivo.

2.3 Stepwise Stochastic Adaptation of a Dictionary

Our approach implements the stepwise stochastic mechanism of gradual improvements introduced in EDA. In this vein, we began by generating ODF dictionaries with the microstructure parameters uniformly distributed in their respective ranges of feasibility (defined in Subsect. 2.1) as suggested in [6]. We will refer to this distribution as *prior* and to such dictionaries as *prior ODF dictionaries*.

For each dataset, we first ran ODF-FP with the respective prior ODF dictionary and looked up the estimated microstructure parameters (Fig. 1a). This first

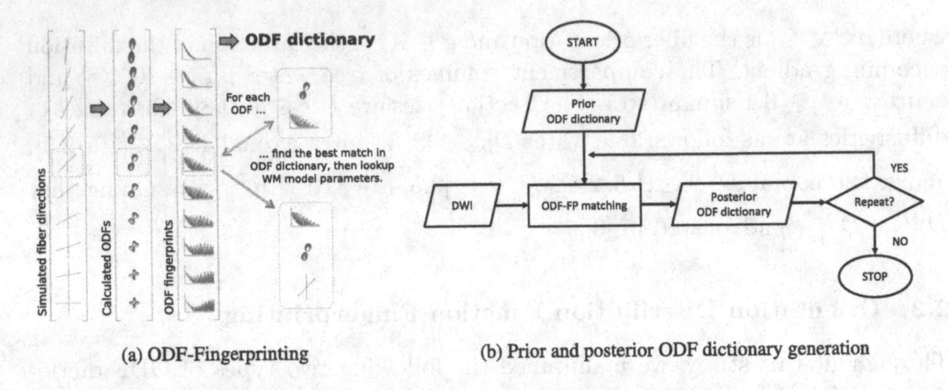

(a) ODF-Fingerprinting (b) Prior and posterior ODF dictionary generation

Fig. 1. Schemes of ODF-FP and the proposed ODF dictionary adaptation mechanism.

run implemented the original ODF-FP procedure (as defined in [2,6]) which we will use for reference.

Later on, we trained two types of Gaussian-based Kernel Density Estimators (KDEs) [19] to represent the empirical distributions of the microstructure parameters that we found with ODF-FP. We defined them as follows:

Type #1 estimator represented the random vector of compartment volumes $(p^{(1)}, \ldots, p^{(N)})$, such that the free water fraction could be computed as $p_{iso} = 1 - \sum_{i=1}^{N} p^{(i)}$. In every experiment, there was only one such estimator.

Type #2 estimators represented the random vectors of diffusivities and intra-axonal volume fractions $(D_{a,\|}^{(i)}, D_{e,\|}^{(i)}, D_{e,\perp}^{(i)}, f^{(i)})$. The number of such estimators depended on the number of distinct sets of fiber parameters per voxel, i.e. one in the phantom dataset and three in vivo.

Based on these, we generated the *posterior ODF dictionary*, such that its elements were no longer uniformly distributed in the space of parameters, but instead they reflected the empirical distribution that we have estimated.

Then, we trained our KDEs again and we used them to generate another instance of the posterior ODF dictionary. We repeated the above procedure 10 times to observe the evolution of the posterior ODF dictionaries in the consecutive iterations of this stepwise adaptation loop (Fig. 1b).

2.4 Data

Diffusion Phantom. We used an anisotropic diffusion phantom manufactured by Psychology Software Tools (Pittsburgh, PA, USA). The phantom contained synthetic fibers made of textile water-filled microtubes (Taxon^{TM} technology [18]) with 0.8 μm diameter. In our experiments, we considered three regions of interest (ROIs) containing pairs of fibers crossing at 90°, 45°, and 30°, as illustrated in Fig. 2.

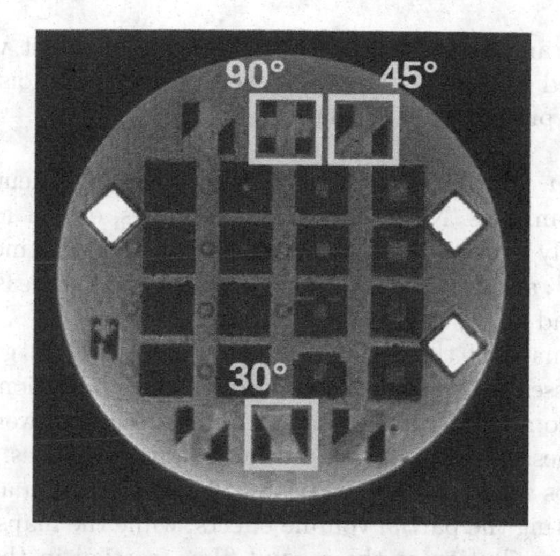

Fig. 2. A slice of the T1-weighted image of the diffusion phantom with 3 regions of interest (in yellow) containing synthetic fibers crossing at 90°, 45°, and 30°. (Color figure online)

We scanned the phantom at 2 mm isotropic resolution, with TE/TR = 74/8000 ms, using a diffusion protocol with 60 exact same sampling directions (forming radial lines [3]) at every b-shell for $b = 1000, 2000, 3000$ s/mm^2, interleaved with 20 images at $b = 0$. We then ran Radial Diffusion Spectrum Imaging (RDSI) [3] to compute ODFs. The MATLAB code that we used for data processing is available at: https://bitbucket.org/sbaete/rdsi_recon

In Vivo Data. We considered one healthy subject from the HCP dataset [21] acquired at 1.25 mm isotropic resolution with $b = 1000, 2000, 3000$ s/mm^2, 90 directions each, interleaved with 18 images at $b = 0$. We computed ODFs for all WM voxels using Generalized Q-sampling Imaging (GQI) [24] pipeline provided in DSI Studio.

2.5 Evaluation

Due to the lack of ground truth values for the microstructure parameters of our diffusion model, we quantified the results by comparing coefficients of variation of the respective variables. To account for the stochastic character of our approach, we repeated every experiment 10 times and computed mean values with standard deviations. Finally, we compared the in vivo results with the values reported in the literature [4,15,22].

3 Results

Our experiments—consisting of 10 iterations of the stepwise stochastic adaptation—were sufficient to observe gradual changes in the estimated microstructure parameters. In many cases, the compartment volumes and diffusivities converged to stable states that were visibly less noisy than the ones obtained with the prior ODF dictionaries.

Diffusion Phantom. Fig. 3 shows the evolution of the coefficients of variation of the estimated parameters. Note that all the variables, except for f_{in}, stabilized after approximately 5 iterations. Among them, the compartment volume fractions (i.e. p_{iso}, $p^{(1)}$, and $p^{(2)}$) increased their dispersion, whereas the diffusivities (i.e. $D_{a,\parallel}$, $D_{e,\parallel}$, and $D_{e,\perp}$) decreased it.

The detailed maps of the estimated parameters (Fig. 4) give more insight into these two classes of convergence. Indeed, as the adaptation mechanism was progressing, the computed compartment volume fractions were evolving from rather blurry images (in the prior case) towards more crisp ones. Particularly, the posterior p_{iso} maps gradually revealed the free water fraction at the boundaries of the fibers reflecting the partial volume effects, while the maps of $p^{(2)}$ correctly highlighted the contribution of the second fiber fraction in the crossing areas. Simultaneously, our maps of diffusivity parameters evolved from fairly scattered images corrupted with noise (in prior ODF dictionary) towards nearly uniform

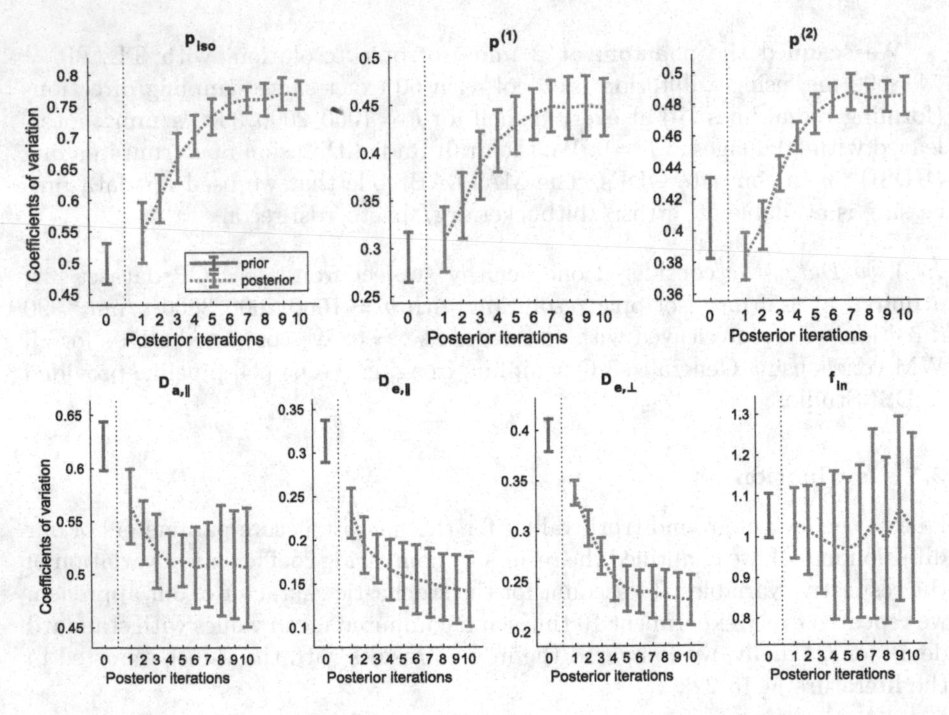

Fig. 3. Coefficients of variations (averaged over 10 runs ± standard deviations) computed on the phantom data converged after approximately 5 iterations in all studied variables except for f_{in}. The plots illustrate the ODF dictionary adaptation process from the prior ODF dictionary (in blue) throughout the 10 iterations of the posterior ODF dictionaries (in red). (Color figure online)

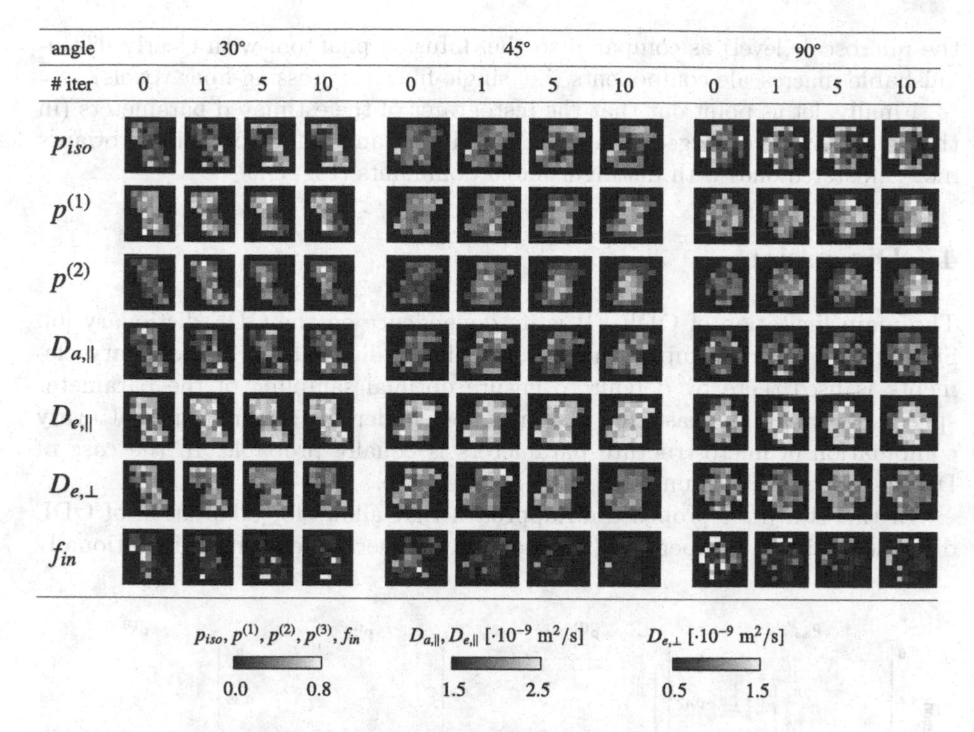

Fig. 4. Detailed maps of the estimated microstructure parameters (in rows) in the diffusion phantom dataset. The images show the adaptation process from the 0th iteration (prior ODF dictionary) throught the 1st, 5th, and 10th iterations of the posterior ODF dictionaries. The regions of interest (in columns) present pairs of synthetic fibers crossing at 30°,45°, and 90°.

maps which better reflected the expected uniform microstructure of the synthetic fibers.

In Vivo Data. We observed a little different convergence pattern on the human brain WM than in the phantom. Here, the ODF dictionary adaptation required more than 5 iterations during which the dispersion changes evolved towards decreasing the coefficients of variation in almost all variables, even the compartment fraction volumes (Fig. 5).

The maps of a sample axial slice (Fig. 6) again provide a more in-depth perspective of the stochastic adaptation process that took place. Note that parameters like p_{iso} and the diffusivities evolved in a similar way to the phantom case, i.e. by emphasizing the partial volume effects (at the boundaries with gray matter or the ventricles) and by smoothing the intra- and extra-axonal diffusivity values. On the other hand, the compartment fraction volumes, especially $p^{(2)}$ and $p^{(3)}$, also tended to decrease their variability (Figs. 5 and 6). This was not surprising due to differences in heterogeneity of the brain tissue (occurring at the microscale level) as compared to the diffusion phantom with clearly distinguishable macroscale components (i.e. single-fiber vs. crossing-fiber voxels).

Finally, let us point out that the histograms of the estimated parameters (in the whole WM) converged from relatively broad and flat distributions towards more clustered ones with distinguishable dominants (Fig. 7).

4 Discussion

The main limitation of ODF-FP is its dependence on the ODF dictionary [6]. Similarly to other lookup techniques, a uniform distribution of dictionary elements is used there by default to ensure optimal sampling of the parameter space. However, this reasoning can only hold under the assumption that every combination of microstructure parameters is equally probable. In the case of DWIs, though, this assumption seems inadequate.

In this study, we proposed an approach that allows for adaptation of ODF dictionaries in an unsupervised, data-driven manner. Moreover, we intentionally

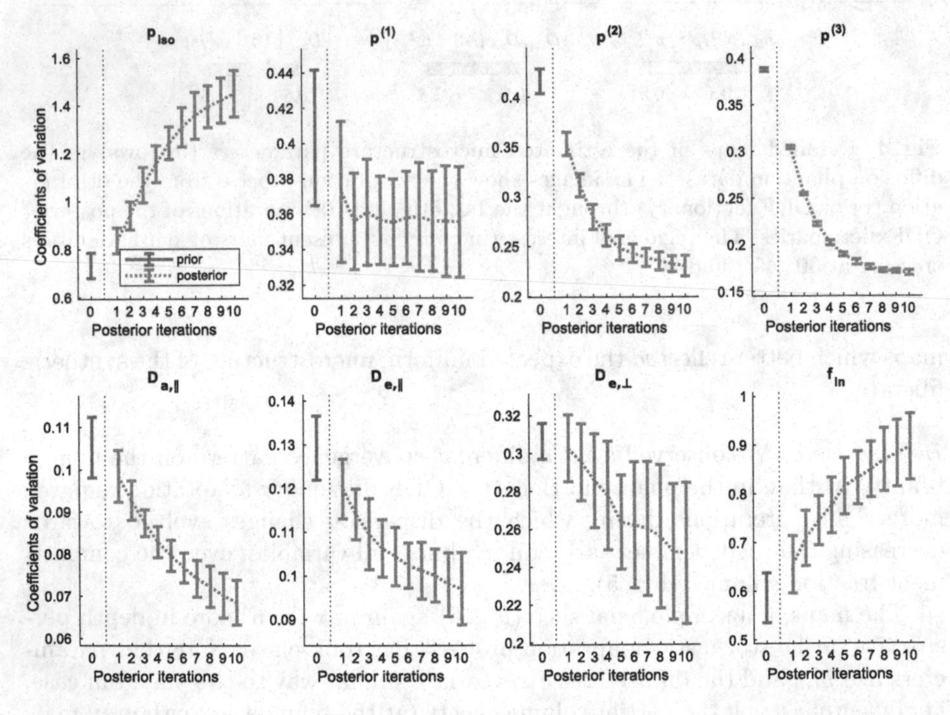

Fig. 5. Coefficients of variations (averaged over 10 runs ± standard deviations) computed on in vivo human data (with the white matter mask applied) were converging towards lower dispersion in all studied variables except for f_{in}. The plots illustrate the ODF dictionary adaptation process from the prior ODF dictionary (in blue) throughout the 10 iterations of the posterior ODF dictionaries (in red). (Color figure online)

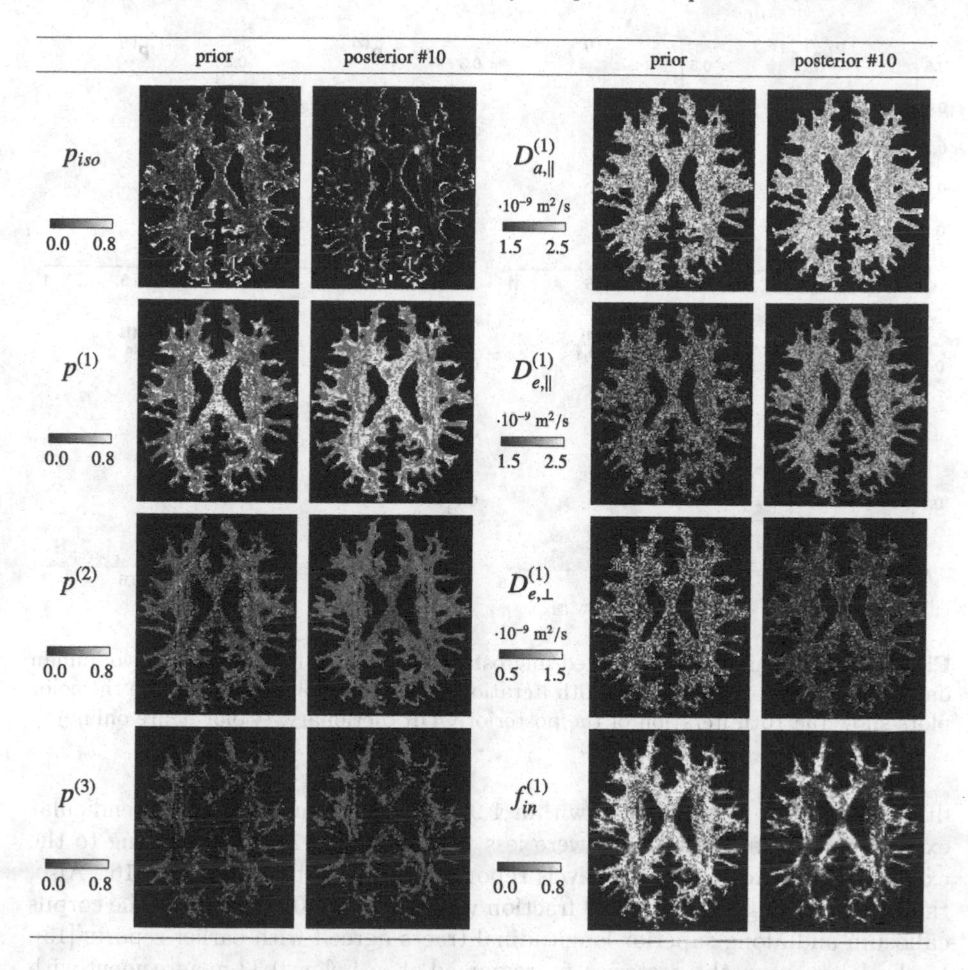

Fig. 6. Detailed maps of the estimated microstructure parameters in the in vivo human dataset. The images show the comparison between the 0th iteration (prior ODF dictionary) and the 10th iteration of the posterior ODF dictionary.

did not impose any extra assumptions on the microstructure parameters (other than the feasibility ranges defined in Subsect. 2.1 and the $D_a \geq D_e$ inequality that were already assumed in ODF-FP [6]) to avoid unwanted bias, e.g. favoring a healthy tissue over pathology. Instead, we simply aimed at replacing a fraction of ODF fingerprints that were highly unlikely to be chosen with the ones that better represented a given dataset. We also required that the algorithm estimates such a distribution automatically.

Our results showed that the values of microstructure parameters that we found with the posterior ODF dictionaries conformed with the values reported in literature [4,15,22]. In WM, most of our reconstructed intra-axonal diffusivities $D_{a,\|}$ ranged between 2.2 and 2.5 $\cdot$ 10^{-9} m^2/s, while the parallel extra-axonal

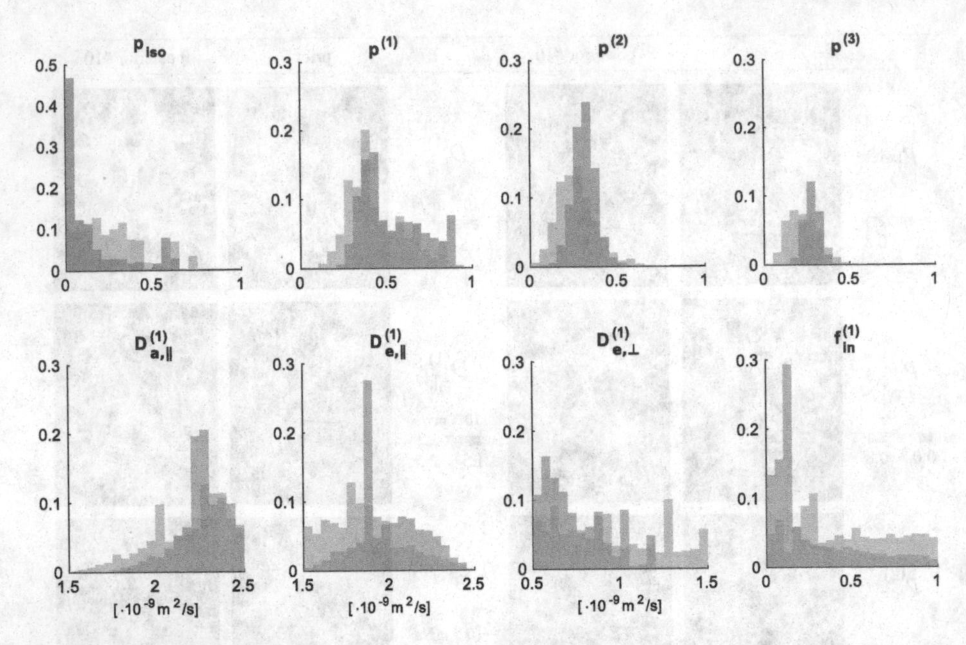

Fig. 7. Histograms of the estimated microstructure parameters in the in vivo human dataset. The gray plots show the 0th iteration (prior ODF dictionary), while the color plots show the 10th iteration of the posterior ODF dictionary. (Color figure online)

diffusivities $D_{e,\|}$ typically lied within 1.9–2.4 $\cdot$ 10^{-9} m^2/s. The perpendicular extra-axonal diffusivities $D_{e,\perp}$ were less than a half of $D_{e,\|}$, conforming to the extra-axonal space tortuosity levels reported in other studies [5,8,10,16]. Also, the clusters of high intra-axonal fraction volumes $f_{in} > 0.6$ located in the corpus callosum and along superior longitudinal tracts agreed with earlier reports [13]. In the other areas, the posterior f_{in} remained at or below 0.33 in agreement with histological findings [8].

Nonetheless, we must point out that our minimal set of assumptions on the microstructure parameters carries a risk of homogenization of the estimated values. The observed tendencies of our approach to smooth diffusivities and volume fractions, especially in the human dataset, or to shift the extra-axonal diffusivities ($D_{e,\|}$ upward and $D_{e,\perp}$ downward) might require a counter-balancing mechanism. Future work should address these issues, for instance, by applying targeted anatomical constraints or DWI noise compensation mechanism.

Moreover, our in vivo study presented in this paper mainly targeted intra-subject reproducibility. In order to draw more general conclusions, a dataset composed of multiple subjects with and without pathologies must be processed next.

5 Conclusions

In this study, we used ODF-FP to estimate the fraction volumes and diffusivities of the multicompartment diffusion model at the linear time complexity. To improve the accuracy of this technique, we proposed a stepwise stochastic adaptation mechanism for generating posterior ODF dictionaries that better reflects the variability of DWIs in hand. As a result, we obtained less noisy estimates of the microstructure parameters and the more pronounced separation of the free water fraction of the diffusion signal.

Acknowledgements. This project was supported in part by the National Institutes of Health (NIH, R01-EB028774 and R01-NS082436). This work was performed under the rubric of the Center for Advanced Imaging Innovation and Research (CAI2R, https://www.cai2r.net), a NIBIB Biomedical Technology Resource Center (NIH P41-EB017183). Some of the data were provided by the Human Connectome Project, WU-Minn Consortium (Principal Investigators: David Van Essen and Kamil Ugurbil; 1U54MH091657) funded by the 16 NIH Institutes and Centers that support the NIH Blueprint for Neuroscience Research; and by the McDonnell Center for Systems Neuroscience at Washington University.

Data Availability Statement. The Python code of ODF-FP with the stepwise stochastic dictionary adaption implemented as an extension of the DIPY library can be downloaded from https://github.com/filipp02/dipy_odffp/tree/odffp

References

1. Akaike, H.: Information theory and an extension of the maximum likelihood principle. In: Proceedings of the 2nd International Symposium on Information, Czaki, Akademiai Kiado, Budapest (1973)
2. Baete, S.H., Cloos, M.A., Lin, Y.C., Placantonakis, D.G., Shepherd, T., Boada, F.E.: Fingerprinting orientation distribution functions in diffusion MRI detects smaller crossing angles. Neuroimage **198**, 231–241 (2019)
3. Baete, S.H., Yutzy, S., Boada, F.E.: Radial q-space sampling for DSI. Magn. Reson. Med. **76**(3), 769–780 (2016)
4. Dhital, B., Reisert, M., Kellner, E., Kiselev, V.G.: Intra-axonal diffusivity in brain white matter. Neuroimage **189**, 543–550 (2019)
5. Fieremans, E., Jensen, J.H., Helpern, J.A.: White matter characterization with diffusional kurtosis imaging. Neuroimage **58**(1), 177–188 (2011)
6. Filipiak, P., Shepherd, T., Lin, Y.C., Placantonakis, D.G., Boada, F.E., Baete, S.H.: Performance of orientation distribution function-fingerprinting with a biophysical multicompartment diffusion model. Magn. Reson. Med. **88**(1), 418–435 (2022)
7. Hauschild, M., Pelikan, M.: An introduction and survey of estimation of distribution algorithms. Swarm Evol. Comput. **1**(3), 111–128 (2011)
8. Jelescu, I.O., Budde, M.D.: Design and validation of diffusion MRI models of white matter. Front. Phys. **5**, 61 (2017)
9. Jelescu, I.O., Palombo, M., Bagnato, F., Schilling, K.G.: Challenges for biophysical modeling of microstructure. J. Neurosci. Methods **344**, 108861 (2020)

10. Jelescu, I.O., Veraart, J., Fieremans, E., Novikov, D.S.: Degeneracy in model parameter estimation for multi-compartmental diffusion in neuronal tissue. NMR Biomed. **29**(1), 33–47 (2016)
11. Jelescu, I.O., et al.: In vivo quantification of demyelination and recovery using compartment-specific diffusion MRI metrics validated by electron microscopy. Neuroimage **132**, 104–114 (2016)
12. Jeurissen, B., Leemans, A., Tournier, J.D., Jones, D.K., Sijbers, J.: Investigating the prevalence of complex fiber configurations in white matter tissue with diffusion magnetic resonance imaging. Hum. Brain Mapp. **34**(11), 2747–2766 (2013)
13. Jung, W., et al.: Whole brain g-ratio mapping using myelin water imaging (MWI) and neurite orientation dispersion and density imaging (NODDI). Neuroimage **182**, 379–388 (2018)
14. Larrañaga, P., Lozano, J.A.: Estimation of distribution algorithms: A new tool for evolutionary computation, vol. 2. Springer Science & Business Media (2001)
15. McKinnon, E.T., Helpern, J.A., Jensen, J.H.: Modeling white matter microstructure with fiber ball imaging. Neuroimage **176**, 11–21 (2018)
16. Novikov, D.S., Fieremans, E.: Relating extracellular diffusivity to cell size distribution and packing density as applied to white matter. In: Proceedings of the 20th Annual Meeting of ISMRM, p. 1829 (2012)
17. Novikov, D.S., Fieremans, E., Jespersen, S.N., Kiselev, V.G.: Quantifying brain microstructure with diffusion MRI: theory and parameter estimation. NMR Biomed. **32**(4), e3998 (2019)
18. Schneider, W., Pathak, S., Wu, Y., Busch, D., Dzikiy, D.J.: Taxon anisotropic phantom delivering human scale parametrically controlled diffusion compartments to advance cross laboratory research and calibration. ISMRM 2019 (2019)
19. Scott, D.W.: Multivariate Density Estimation: Theory, Practice, and Visualization. John Wiley & Sons (2015)
20. Taquet, M., et al.: Extra-axonal restricted diffusion as an in-vivo marker of reactive microglia. Sci. Rep. **9**(1), 1–10 (2019)
21. Van Essen, D.C., et al.: The human connectome project: a data acquisition perspective. Neuroimage **62**(4), 2222–2231 (2012)
22. Veraart, J., Fieremans, E., Novikov, D.S.: On the scaling behavior of water diffusion in human brain white matter. Neuroimage **185**, 379–387 (2019)
23. Wedeen, V.J., et al.: Diffusion spectrum magnetic resonance imaging (DSI) tractography of crossing fibers. Neuroimage **41**(4), 1267–1277 (2008)
24. Yeh, F.C., Wedeen, V.J., Tseng, W.Y.I.: Generalized q-sampling imaging. IEEE Trans. Med. Imaging **29**(9), 1626–1635 (2010)

How Can Spherical CNNs Benefit ML-Based Diffusion MRI Parameter Estimation?

Tobias Goodwin-Allcock[1](✉), Jason McEwen[2], Robert Gray[3], Parashkev Nachev[3], and Hui Zhang[1]

[1] Department of Computer Science and Centre for Medical Image Computing, University College London, London, UK
tobias.goodwin-allcock@ucl.ac.uk
[2] Kagenova Limited, Guildford, UK
[3] Department of Brain Repair and Rehabilitation, Institute of Neurology, UCL, London, UK

Abstract. This paper demonstrates spherical convolutional neural networks (S-CNN) offer distinct advantages over conventional fully-connected networks (FCN) at estimating scalar parameters of tissue microstructure from diffusion MRI (dMRI). Such microstructure parameters are valuable for identifying pathology and quantifying its extent. However, current clinical practice commonly acquires dMRI data consisting of only 6 diffusion weighted images (DWIs), limiting the accuracy and precision of estimated microstructure indices. Machine learning (ML) has been proposed to address this challenge. However, existing ML-based methods are not robust to differing gradient schemes, nor are they rotation equivariant. Lack of robustness to differing gradient schemes requires a new network to be trained for each scheme, complicating the analysis of data from multiple sources. A possible consequence of the lack of rotational equivariance is that the training dataset must contain a diverse range of microstucture orientations. Here, we show spherical CNNs represent a compelling alternative that is robust to new gradient schemes as well as offering rotational equivariance. We show the latter can be leveraged to decrease the number of training datapoints required.

Keywords: Spherical CNNs · Machine learning · Diffusion MRI

1 Introduction

Diffusion MRI (dMRI) plays an important role in neuroscientific and clinical research because it can help infer tissue microstructure [3]. To infer tissue microstructure from dMRI, we use mathematical models to estimate parameters from dMRI data. At each voxel, the measurements made according to some

This work is supported by the EPSRC-funded UCL Centre for Doctoral Training in Medical Imaging (EP/L016478/1), the Department of Health's NIHR-funded Biomedical Research Centre at UCLH and the Wellcome Trust.

S. Cetin-Karayumak et al. (Eds.): CDMRI 2022, LNCS 13722, pp. 101–112, 2022.
https://doi.org/10.1007/978-3-031-21206-2_9

acquisition scheme - commonly consisting of gradient schemes coupled with their diffusion sensitising factor - are fitted to mathematical models, such as the diffusion tensor (DT) [2]. From these models, dMRI parameters can be derived to reveal the microstructure such as fractional anisotropy (FA), which characterises the anisotropy of the tissue. The computation of these parameters from dMRI data is known as dMRI parameter estimation, which has traditionally been achieved with model fitting. However, the fidelity of dMRI parameters estimated in this way is limited by relatively high noise in the data, requiring more measurements to be acquired than what are routinely made in the clinic [10].

As in many other fields, dMRI parameter estimation has recently been revolutionised by exploiting deep learning (DL), yielding greatly increased accuracy than conventional fitting when the acquisition scheme has a small number of samples [1,8]. However, current deep-learning methods, e.g. fully-connected networks (FCN), are ignorant of the acquisition scheme of a given acquisition, rendering these methods potentially not generalisable to new acquisition schemes. This complicates the application of a DL model to data acquired from multiple sources. Moreover, these methods do not exhibit rotational equivariance, a property that may help reduce the demand for training data.

There have been a number of attempts to capture the relationship between an acquisition scheme and the corresponding data [4,12]. However, they do not utilise the topological features of the associated gradient schemes. Gradient schemes for a given diffusion sensitising factor can be represented by points on the unit sphere. Therefore, spherical convolutional neural networks (S-CNN), recently proposed as an alternative to FCNs [7,13], provide a more natural solution to this problem. However, currently there exists no direct evidence of the theoretical benefits of S-CNNs, such as rotational equivariance and robustness to different gradient schemes. Here we aim to provide the very first empirical evidence of these advantages in the context of estimating rotation-invariant dMRI parameters.

The rest of the paper is described as follows: Sect. 2 how machine learning has been used to solve the dMRI parameter estimation problem and the theoretical beneficial properties of S-CNNs; Sect. 3 then goes on to explain how we empirically test these properties; Sect. 4 summarises the results and discusses future work.

2 ML Solutions to the dMRI Parameter Estimation Problem and the Theoretical Benefits of S-CNNs

Deep learning (DL) has been proposed as a solution to dMRI parameter estimation from small numbers of diffusion weighted images (DWI). This section provides (1) an example of the current machine learning standard for voxel-wise estimation (2) the theoretical limitations of this architecture (3) the architecture features that theoretically benefit S-CNNs. As an example, we show how the dMRI parameter FA is estimated from a common clinical diffusion MRI acquisition consisting of 6 DWIs.

Deep learning models, F, map the dMRI signals, $\mathbf{s}$, and their corresponding acquisition scheme - consisting of b-values, $\mathbf{B}$, and gradient scheme $\mathbf{G}$ - directly to dMRI parameters, denoted t.

$$t = F(\mathbf{s}, \mathbf{B}, \mathbf{G}; \theta) \tag{1}$$

This function is learnt by optimising the trainable parameters, θ, on training data. After training, the quality of the network estimation depends on many factors. The two factors explored in this work are the training data distribution and the choice of network architecture.

2.1 Fully-Connected Networks

The first and most common deep learning network architecture applied to dMRI data is the FCN [1,8]. Conventionally these have been implemented following:

$$t = F_{\text{FCN}}(\mathbf{s}; \theta_{FCN}), \tag{2}$$

where F_{FCN} is a fully-connected network with trainable parameters θ_{FCN}. The network's input consists of the dMRI signals. Absent from the equation is the acquisition scheme so the network is ignorant of the acquisition scheme. Estimation from a new set of DWIs is accurate only if the acquisition scheme for the new data is consistent with the acquisition scheme during training [12].

An FCN's architecture is not designed to be rotationally equivariant. A theoretical consequence of lacking rotational equivariance is that the training dataset may have to contain a diverse set of tissue microstructure orientations for FCNs to accurately estimate independent of fibre orientation.

2.2 Spherical CNNs

S-CNNs theoretically improve over FCNs - both in terms of robustness to the gradient scheme and robustness to the training data distribution - because of the difference in network architecture.

An S-CNN's architecture differs greatly to an FCN's but not in the way one may expect. In S-CNNs, the convolution isn't across multiple voxels, like traditional CNNs, but over the spherical image space. Therefore, S-CNNs are voxel wise networks just like FCNs. The spherical image is generated at each voxel from the dMRI signals, $\mathbf{s}$, along with their corresponding gradient scheme $\mathbf{G}$. We see that this architecture may naturally address the highlighted limitations of FCNs. Firstly, an S-CNN's input is informed of the gradient scheme as shown in the following equation:

$$t = F_{\text{S-CNN}}(\mathbf{s}, \mathbf{G}; \theta_{\text{S-CNN}}) \tag{3}$$

where $F_{\text{S-CNN}}$ is an S-CNN with trainable parameters $\theta_{\text{S-CNN}}$. We hypothesise this input will allow S-CNNs to be robust to a change in gradient scheme at inference time as long as the diffusion sensitising factors are the same.

Another benefit of S-CNN's is the rotationally equivariant architecture [6]. We hypothesise that this property will allow S-CNNs to extrapolate from a training dataset with a common primary fibre orientation and, during the inference stage, well estimate tissue with fibres oriented along any direction. As a result, S-CNNs do not require a diverse set of tissue microstructure orientations in the training dataset, reducing demands on the training dataset.

3 Experiments

Each claim made in this paper is evaluated with an individual experiment. The first experiment evaluates network robustness to differing gradient schemes; the second assesses network robustness to the distribution of the primary fibre orientations in the training set.

3.1 Experiment 1

Study Design. In this experiment we test if the networks are robust to new gradient schemes at inference time. In order to show this we propose an experiment where both networks are trained with a typical gradient scheme and then, in the inference phase, the trained networks are applied to data collected using a gradient scheme (1) the same as the training gradient scheme and another (2) different to the training scheme.

Network Architectures and Training Parameters. Both network architectures are voxel-wise networks. The S-CNN architecture we use, known as the hybrid spherical CNN architecture [5], was chosen as it has been shown to be highly rotationally equivariant whilst also being computationally efficient. The specific network parameters follow the spherical MNIST experiment and the input to this network is a densely sampled spherical signal, described later. The FCN network architecture implementation, used as the baseline, is consistent with the established implemented FCN techniques for dMRI parameter estimation [1,8]. This network input follows the standard practice consisting of 6 $b = 0$ normalised diffusion-weighted signals. The network hyperparameters are: three hidden layers with number of units = [100,100,10] and ReLU activation function.

The training parameters are chosen in order for FCN to perform optimally and consistent between the networks so that any difference between trained networks is solely because of their architectures. To achieve this, the training parameters are consistent with the FCN literature; specifically the training regime used Adam optimiser for 50 epochs with learning rate set to 0.001, the batch size 32 and the loss metric MSE.

Generating Densely Sampled Spherical Signals. S-CNNs require densely sampled spherical signals as input. Densely sampled spherical signals are generated for each voxel by utilising the property of the 1-to-1 mapping between six-directional dMRI signals and the 6 independent values of the diffusion tensor. Due to this property, six-directional dMRI signals, with all of their noise, are perfectly and uniquely described by a DT. From this DT a spherical function called the ADC profile may be derived and sampled to generate the input required for S-CNNs. This process is visually described in Fig. 1.

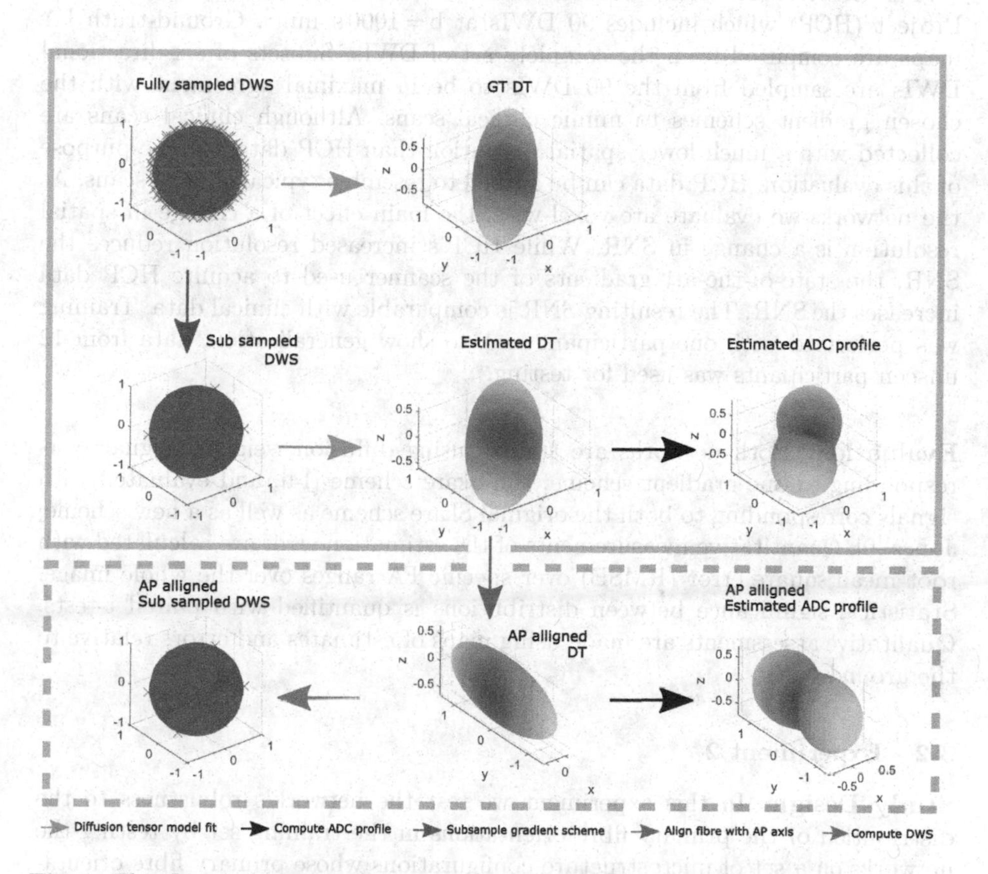

Fig. 1. This figure shows how the testing data is generated for the both experiments (blue box). The figure also shows how the training data is generated for experiment 1 (blue box) and experiment 2 (green box). In the blue box, from the 90 directional $b = 0$ normalised diffusion weighted signals (DWS) we estimate the ground truth diffusion tensor, and generate a 6-dir clinical scan by subsampling. The clinical scan's measurements are used as input to the FCN, but for compatibility with the S-CNN they must first be re-parameterised as an ADC profile. This is achieved by exploiting the 1-to-1 relationship described in Sect. 3.1. For the second experiment training data is required that has the primary fibre oriented along the AP axis. This is achieved by changing the orientation of the estimated DT and calculating the 6 DWS, for FCN, or calculating the ADC profile for S-CNN. (Color figure online)

Datasets. A dataset is required for training and testing the models. All of the deep learning models evaluated in this paper are supervised machine learning techniques, therefore, the training dataset must consist of a set of input values paired with ground truth output values. For the high-quality ground truth output, a dataset is required that contains a sufficiently large number of DWIs to provide accurate estimation of FA.

For this reason, we have chosen dMRI data from the Human Connectome Project (HCP) which includes 90 DWIs at $b = 1000\,s/mm^2$. Ground-truth FA maps are computed from the complete set of DWIs. Subsets of six-directional DWIs are sampled from the 90 DWIs to be in maximal agreement with the chosen gradient schemes to mimic clinical scans. Although clinical scans are collected with a much lower spatial resolution than HCP data, for the purpose of this evaluation, HCP data can be argued to resemble typical clinical scans. As the networks we evaluate are voxel-wise, the main effect of a change in spatial resolution is a change in SNR. While HCP's increased resolution reduces the SNR, the state-of-the-art gradients of the scanner used to acquire HCP data increases the SNR. The resulting SNR is comparable with clinical data. Training was performed with one participant and, to show generalisation, data from 12 unseen participants was used for testing.

Evaluation. Both networks are trained using diffusion weighted signals corresponding to one gradient scheme, the Skare scheme [14], and evaluated with signals corresponding to both the original Skare scheme as well as a new scheme, Jones [9]. Quantitative measurements of the estimation error are calculated with root mean square error (RMSE) over specific FA ranges over the whole image. Statistical significance between distributions is quantified with paired t-tests. Qualitative assessments are made using maps of estimates and errors relative to the ground truth.

3.2 Experiment 2

Study Design. In this experiment we test the network's robustness to the distribution of the primary fibre orientations in the training set by testing the networks on a set of microstructure configurations whose primary fibre orientation lies both inside and outside of the training dataset's distribution. We achieve this by restricting the primary fibre orientation in the training dataset to align with the anterior-posterior axis and testing on microstructure oriented in all directions. We hypothesise that networks robust to the distribution of primary fibre orientation in the training set will estimate the FA equally well independent of the direction of the primary fibre orientation.

Restricting Primary Fibre Orientation. The training distribution of the primary fibre orientations is restricted to the anterior-posterior axis by adapting the dense ADC sampling algorithm. In the dense ADC sampling algorithm, after a noisy DT's estimation, the noisy DT's shape and size are extracted by eigendecomposition. Next, the primary fibre orientation, otherwise known as the principal eigenvector, is set to the anterior-posterior axis whilst the secondary eigenvector is set to the superior-inferior axis; see visual description in the light green dashed box in Fig. 1. This new AP-aligned diffusion tensor is used to generate the input required for FCN, by computing the 6 directional $b = 0$ normalised diffusion-weighted signals from the DT forward model, and the input required for S-CNN, by densely sampling the ADC profile.

Evaluation. Network architectures, training scheme, training dataset and evaluation metrics are the same as experiment 1. Only one gradient scheme is required for this experiment so the Skare scheme is chosen for training and testing. Training data undergoes primary fibre orientation restriction, whilst testing data is unrestricted. To show further benefits of rotational equivariance we test to see if this property allows S-CNNs to estimate with high fidelity when starved of training data points. For this, an S-CNN is trained with only 10 % of the total training datapoints. The distribution of estimation error over the primary fibre orientation is evaluated.

4 Results and Discussion

4.1 For Experiment 1

Qualitative and quantitative results of experiment 1 are shown in Figs. 2 and 3 respectively.

Figure 2 shows an example slice of a GT FA map from a test subject along with the model fitting, FCN and S-CNN estimations and error maps when the gradient scheme is the same or different between training and testing. When the training and testing schemes are the same FCN's performance is consistent with the literature, estimating FA faithfully. When the gradient schemes are different the FCN estimates poorly. Estimation is especially poor in areas of high FA, such as the corpus callosum. S-CNN's estimations are similar to the GT regardless of the gradient scheme. This shows S-CNN's robustness to differing gradient schemes.

Figure 3 reinforces the qualitative observations with quantitative measures showing boxplots of the mean RMSE over the 12 subjects. We see for anisotropic signals ($FA > 0.4$) the conventional FCN performance is significantly worse ($p < 8e-12$) when applied to a new gradient scheme than on the training scheme, whereas, the S-CNN models estimation fidelity does not decrease when applied to a new scheme. In both figures, on differing gradient schemes the FCN model is shown to be more inaccurate in regions of high FA. This loss of accuracy in high FA regions may be caused by the signal attenuation in these regions being greatly dependant on the direction of measurement. Therefore, for low FA voxels the DWS from any two gradient schemes are similar. Whereas, for areas of high FA the DWS will be vastly different and therefore the training distribution will be different to the testing distribution.

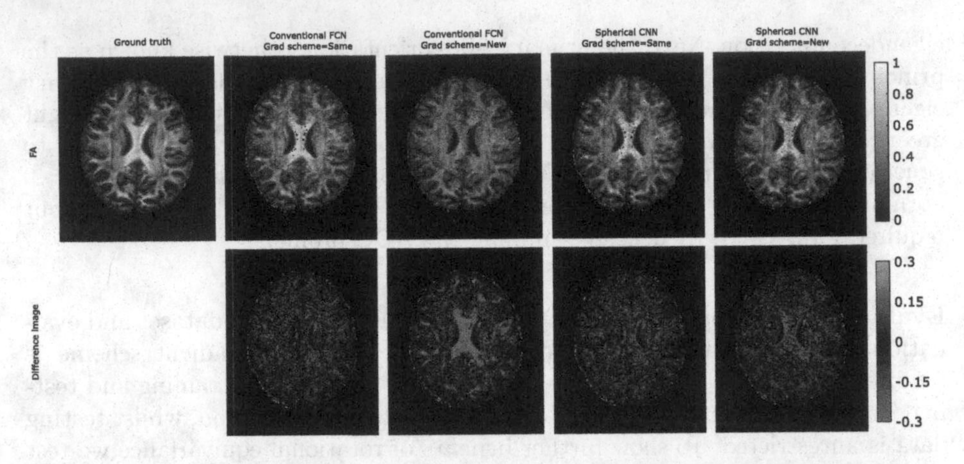

Fig. 2. Results of experiment 1 are visualised on an subject unseen during training. Conventional fully-connected networks are compared against the S-CNN model at estimating FA with either the same gradient scheme as used in training or a new gradient scheme. The FCN shows a drop in performance when the gradient scheme is different between training and testing. The S-CNN doesn't have this issue, therefore, it is robust to differing gradient schemes.

Gradient scheme robustness may also be achieved with FCN by first encoding the signals in terms of their corresponding spherical harmonic representation [11]. This approach is essentially equivalent to S-CNN but does not offer the property of rotational equivariance, the effect of which is demonstrated in the next subsection.

4.2 For Experiment 2

The results of experiment 2 are shown in Figs. 3, 4 and 5. Figure 4 qualitatively shows the effect of estimating the full brain volume using networks trained only on tissue microstructure aligned with the anterior-posterior axis. The FCN model consistently underestimates FA in regions where the underlying tissue microstructure does not align with anterior-posterior direction (e.g. the corpus callosum which consists of left-right white matter tracts). In contrast, the S-CNN models accurately estimate FA independently of the primary fibre direction, and the error is far less structured than the FCN's.

This is mirrored in the quantitative measurements over the 12 testing subjects, shown in Fig. 3. The FCN model only well estimates isotropic signals (FA < 0.2); however the performance difference between different methods is relatively small, with the difference between the mean RMSE for different methods no larger than 0.04. When the signal becomes anisotropic, the FCN model estimates the signal significantly worse than the S-CNN models, with the difference in the mean RMSE for different methods no smaller than 0.08. As the signals get more anisotropic the FCN model performs even worse until in the top FA bracket

($0.8 \leq \text{FA} < 1$) the mean RMSE between the FCN and the S-CNN models is 0.5, 10 times the difference in error of the isotropic signals.

Figure 5 shows the distribution of the absolute error over the full range of primary fibre orientations for anisotropic signals. The FCN model well estimates tissue microstructure aligned with the anterior-posterior axis, seen during training. However the error quickly grows as the primary fibre orientation deviates from this axis. This adverse feature is not exhibited by the S-CNN model as the estimation error is low and independent of training dataset distribution of the primary fibre orientation. The lack of rotational equivariance in FCNs hinders estimation performance when generalising to microstructure with primary fibre orientation not sampled in the training distribution. This has potential for orien-

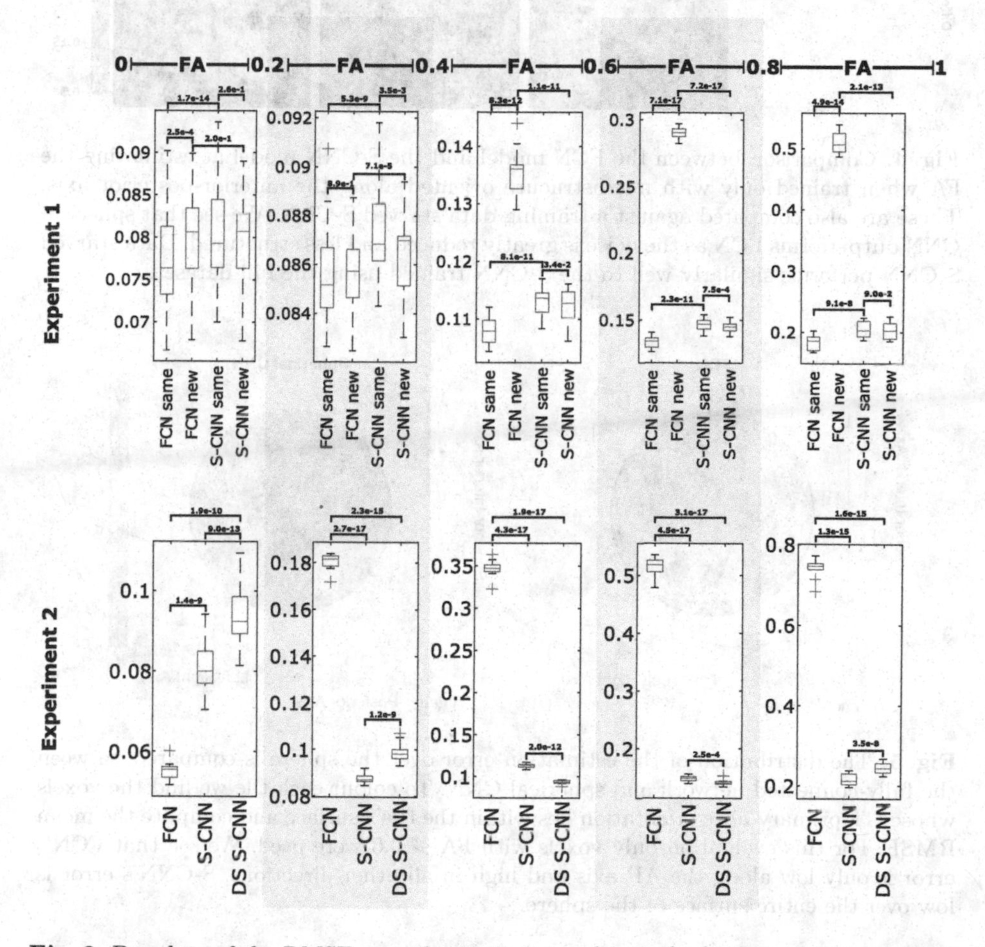

Fig. 3. Boxplots of the RMSE over the 12 testing subjects for both experiments, with p-values between distributions from paired t-test shown. The testing data is uniformly split into 5 GT FA ranges. For both experiments the problems with FCN become apparent when applied to anisotropic signal.

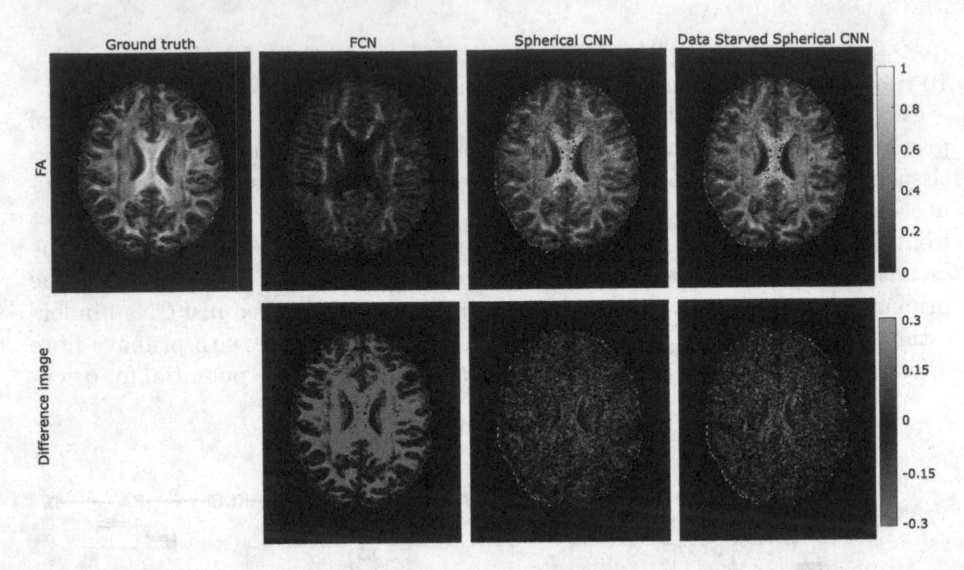

Fig. 4. Comparison between the FCN model and the S-CNN model at estimating the FA when trained only with microstrucure oriented along the anterior-posterior axis. These are also compared against a training data starved S-CNN. We see that spherical CNN outperforms FCN as the noise is greatly reduced and less structured. Data starved S-CNN performs similarly well to the S-CNN trained using the full dataset.

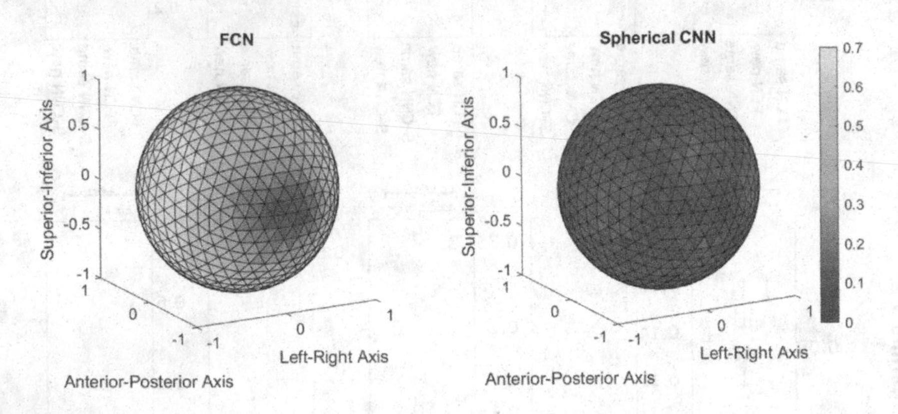

Fig. 5. The distribution of the estimation error over the sphere is compared between the fully-connected network and spherical CNN. To colour each tile we find the voxels whose GT primary fibre orientation lies within the tile's surface and compute the mean RMSE. For this evaluation only voxels with FA > 0.6 were used. We see that FCN's error is only low along the AP axis and high in all other directions. S-CNN's error is low over the entire surface of the sphere.

tation bias in the training dataset to lead to poor estimation for under-sampled directions.

In Figs. 3 and 4 another advantage of rotational equivariance is shown. Performance of the S-CNN network is not greatly changed when only a tenth of the training dataset is used. This is shown in the maps of FA estimated by the data starved S-CNN model in Fig. 4 and the RMSE over the 12 testing subjects shown in Fig. 3. The robustness to a lack of training data is due to S-CNNs being rotationally equivariant. Therefore, during training only a diverse distribution of the microstructure shape is required and not their orientations as well. This property of S-CNNs is a real benefit as it reduces the number of training datapoints required for good estimation at inference stage and, when simulating training data, allows for denser sampling of the shape parameters as SO(3) need not be sampled.

5 Conclusion

In this work we explore the advantages of S-CNNs for dMRI parameter estimation over conventional FCNs. Representing diffusion weighted signals as a spherical image is here demonstrated to gain robustness to the gradient scheme absent from conventional FCNs, at no cost to fidelity. This removes the need to retrain a new network for every gradient scheme, a feature especially beneficial when combining data from multiple sites. S-CNN is shown to be superior to FCN methods additionally because of its rotational equivariance property. This enables the network to encode information about the pattern of the signal irrespective of primary fibre orientation. This eliminates the need to sample diffusion primary fibre orientations, reducing the number of samples needed to cover the full parameter space.

References

1. Aliotta, E., Nourzadeh, H., Sanders, J., Muller, D., Ennis, D.B.: Highly accelerated, model-free diffusion tensor MRI reconstruction using neural networks. Med. Phys. **46**(4), 1581–1591 (2019). https://doi.org/10.1002/mp.13400
2. Basser, P.J., Mattiello, J., LeBihan, D.: Estimation of the effective self-diffusion tensor from the NMR spin echo. J. Magn. Resonan. Ser. B **103**(3), 247–254 (1994). https://doi.org/10.1006/jmrb.1994.1037
3. Bodini, B., Ciccarelli, O.: Diffusion MRI in Neurological Disorders. Diffusion MRI: From Quantitative Measurement to In vivo Neuroanatomy, 2nd edn, pp. 241–255 (2014). https://doi.org/10.1016/B978-0-12-396460-1.00011-1
4. Chen, G., et al.: Estimating tissue microstructure with undersampled diffusion data via graph convolutional neural networks. In: Martel, A.L., et al. (eds.) MICCAI 2020. LNCS, vol. 12267, pp. 280–290. Springer, Cham (2020). https://doi.org/10.1007/978-3-030-59728-3_28
5. Cobb, O.J., et al.: Efficient generalized spherical CNNs. In: ICLR 2021 (2021). https://arxiv.org/abs/2010.11661

6. Cohen, T.S., Geiger, M., Köhler, J., Welling, M.: Spherical CNNs. In: ICLR 2018, January 2018. https://arxiv.org/abs/1801.10130
7. Elaldi, A., Dey, N., Kim, H., Gerig, G.: Equivariant spherical deconvolution: learning sparse orientation distribution functions from spherical data. In: Feragen, A., Sommer, S., Schnabel, J., Nielsen, M. (eds.) IPMI 2021. LNCS, vol. 12729, pp. 267–278. Springer, Cham (2021). https://doi.org/10.1007/978-3-030-78191-0_21
8. Golkov, V., et al.: q-Space deep learning: twelve-fold shorter and model-free diffusion MRI scans. IEEE Trans. Med. Imaging **35**(5), 1344–1351 (2016). https://doi.org/10.1109/TMI.2016.2551324
9. Jones, D.K., Horsfield, M.A., Simmons, A.: Optimal strategies for measuring diffusion in anisotropic systems by magnetic resonance imaging. Magn. Reson. Med. **42**(3), 515–525 (1999). https://doi.org/10.1002/(SICI)1522-2594(199909)42:3⟨515::AID-MRM14⟩3.0.CO;2-Q
10. Jones, D.K.: The effect of gradient sampling schemes on measures derived from diffusion tensor MRI: a Monte Carlo study. Magn. Reson. Med. **51**(4), 807–815 (2004). https://doi.org/10.1002/mrm.20033
11. Lin, Z., et al.: Fast learning of fiber orientation distribution function for MR tractography using convolutional neural network. Med. Phys. **46**(7), 3101–3116, mp.13555 (2019). https://doi.org/10.1002/mp.13555
12. Park, J., et al.: DIFFnet: diffusion parameter mapping network generalized for input diffusion gradient schemes and b-values. IEEE Trans. Med. Imaging **41**(2), 491–499 (2022). https://doi.org/10.1109/TMI.2021.3116298
13. Sedlar, S., Alimi, A., Papadopoulo, T., Deriche, R., Deslauriers-Gauthier, S.: A spherical convolutional neural network for white matter structure imaging via dMRI. In: de Bruijne, M., et al. (eds.) MICCAI 2021. LNCS, vol. 12903, pp. 529–539. Springer, Cham (2021). https://doi.org/10.1007/978-3-030-87199-4_50
14. Skare, S., Hedehus, M., Moseley, M.E., Li, T.Q.: Condition number as a measure of noise performance of diffusion tensor data acquisition schemes with MRI. J. Magn. Resonan. **147**(2), 340–352 (2000). https://doi.org/10.1006/jmre.2000.2209

Tractography and WM Pathways

DC²U-Net: Tract Segmentation in Brain White Matter Using Dense Criss-Cross U-Net

Haoran Yin[1], Pengbo Xu[1], Hui Cui[3], Geng Chen[2(✉)], and Jiquan Ma[1(✉)]

[1] Department of Computer Science and Technology, Heilongjiang University, Harbin, China
majiquan@hlju.edu.cn
[2] National Engineering Laboratory for Integrated Aero-Space-Ground-Ocean Big Data Application Technology, Northwestern Polytechnical University, Xi'an, China
geng.chen@nwpu.edu.cn
[3] La Trobe University, Melbourne, Australia

Abstract. Diffusion magnetic resonance imaging (dMRI) is a non-invasive technique for studying the microstructure properties of brain white matter (WM) in vivo. Segmentation of WM fiber tracts can be used to characterize the topological structure of the human brain and to exploit the biological landmark of abnormal areas by dMRI. To improve the performance of the fiber tract segmentation, we propose a novel U-Net based architecture with dense criss-cross attention, which captures non-local rich global contextual information more efficiently. Our model is evaluated using the real brain data from Human Connectome Project (HCP). Extensive experiments demonstrate that our model improves the performance of fiber tract segmentation, especially for the fiber bundle with complicated topology structure.

Keywords: Diffusion MRI · Fiber tract segmentation · Attention · Dense connection

1 Introduction

Diffusion magnetic resonance imaging (dMRI), as a unique non-invasive method, has facilitated tremendous progress in the study of the microstructure of the human brain. White matter (WM) fiber tracts connect different regions of the gray matter, allowing information transmission in the whole brain. Brain diseases have been found to be closely associated with morphological changes in specific WM tracts [1,23]. Therefore, how to improve the segmentation accuracy of WM fiber tracts is very critical for neuroscience studies and brain disorder diagnosis.

Efforts have been dedicated to accurate WM tract segmentation. For instance, Wassermann et al. [21] proposed a region-of-interest (ROI) based approach to classify fiber tracts by processing fiber streamlines. Garyfallidis et al.

This work was supported in part by the Natural Science Foundation of Heilongjiang Province under Grant LH2021F046.

S. Cetin-Karayumak et al. (Eds.): CDMRI 2022, LNCS 13722, pp. 115–124, 2022.
https://doi.org/10.1007/978-3-031-21206-2_10

[6] used a clustering-based approach to classify and select reference fiber streamlines. Recently, deep learning techniques have been widely used in WM fiber tract segmentation. Li et al. [12] proposed a 3D convolutional neural network (CNN) based approach for WM fiber tract segmentation and performed validation on a large-scale dataset. Liu et al. [13] developed a graph convolutional neural network (GCNN) to predict the grouping labels of individual fiber bundles. Lu et al. [14] employed an encoder-decoder CNN to address data annotation scarcity in WM region segmentation. Wasserthal et al. [22] calculated and embedded fiber orientation maps to a 2D FCNN.

Despite the progress in tract segmentation, existing works usually rely on simple fully convolutional networks (FCNs), such as U-Net. FCNs are effective in capturing the local context information, but are unable to fully locate the long-range dependencies, which are crucial to accurate tract segmentation since the global context information provides valuable high-level semantic clues that are the keys to identifying tract bundles. The self-attention mechanism is proposed to exploit the long-range dependencies with powerful non-local operations. It has been demonstrated to be effective in segmentation tasks [16], but suffers from a limitation of large computational burden, which restricts its wide applications. Furthermore, it has not been investigated in the tract segmentation task. Therefore, an FCN equipped with an efficient self-attention module is greatly desired for accurately segmenting WM tracts.

To this end, we proposed a dense criss-cross U-Net (**DC^{2}U-Net**) to accurately segment tract bundles from fiber peaks. On the basis of U-Net [18], our DC^{2}U-Net innovatively integrates an efficient self-attention component, criss-cross attention [9], into the model, which improves the robustness and performance of fiber tract segmentation. In addition, we add dense connections to optimize the original criss-cross attention, yielding an improved self-attention module, called dense criss-cross attention (DCCA). In general, DC^{2}U-Net uses DCCA to capture richer global context information and employs the resulting context information to improve the accuracy of tract segmentation. Finally, we design a loss function with deep supervision to improve the training procedure. Our DC^{2}U-Net segments brain WM fiber tracts from fiber peaks rather than raw dMRI data, allowing better generalization ability to work with dMRI data acquired using different sampling schemes in q-space. We evaluate the proposed model using the public Human Connectome Project (HCP) dataset [5]. Quantitative and qualitative comparisons demonstrate that DC^{2}U-Net improves the WM fiber tract segmentation performance effectively.

2 Methods

2.1 Dense Criss-Cross U-Net

Conventional automatic segmentation of WM fiber tracts depends on either fiber clustering or ROI-based methods, which are difficult to make a joint optimization and consume a lot of computational resources, therefore they are difficult to be applied in the practical clinic.

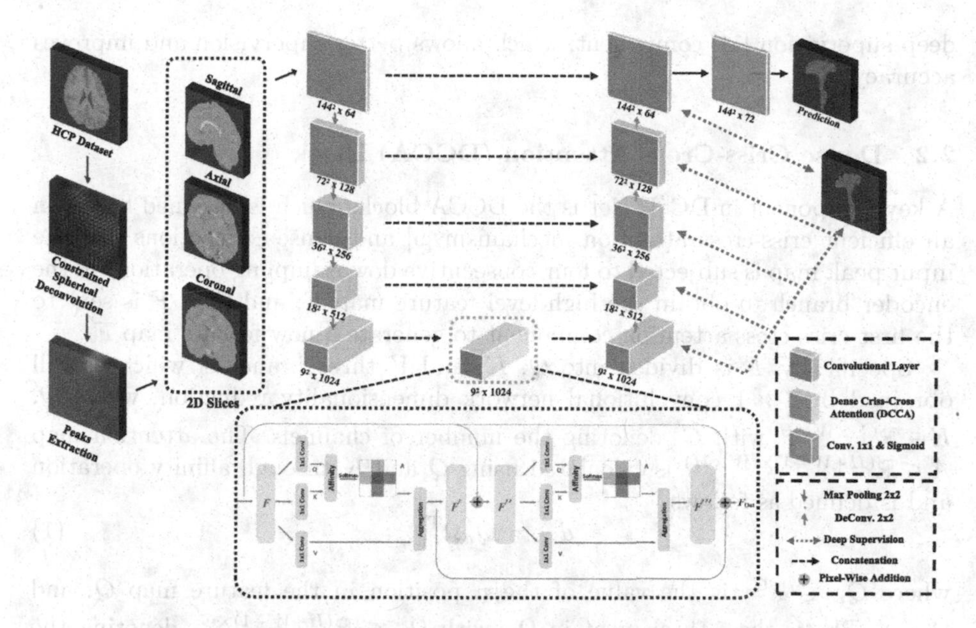

Fig. 1. The overall architecture of DC^{2}U-Net. Raw HCP data is firstly processed by constrained spherical deconvolution and peaks extraction. Afterward, we fed random inputs from three directions (coronal/axial/sagittal) into a 2D U-Net encoder for feature extraction. The resulting features are then sent to the DCCA block for feature enhancement. Finally, deep features are fed to the decoder for predicting the segmentation maps.

Different from conventional methods, our model segments fiber bundles directly without additional operations such as fiber clustering or ROI labeling. Instead of using raw dMRI data as network input, we use fiber peaks instead for better model generalization, as in [22]. For this purpose, we use multi-shell multi-tissue constrained spherical deconvolution (CSD) [10] to compute fODF and then extract a maximum number of three peaks per voxel.

The proposed DC^{2}U-Net consists of three major components, including a segmentation encoder, a dense criss-cross attention (DCCA) block, and a decoder. As shown in Fig. 1, random inputs from different directions (coronal/axial/sagittal) are firstly fed into the encoder. The feature maps generated by four consecutive max pooling operations are sent to the proposed DCCA to extract enhanced context information. Finally, deep features are reconstructed using four successive learnable deconvolutions for up-sampling operations.

As in [22], during the training, we randomly select an oriented (coronal/axial/sagittal) slice from one subject and feed it into the network to predict the probability map for each fiber bundle. Subsequently, these maps corresponding to different orientations are merged by averaging during the test to produce the final 3D segmentation results. Furthermore, we trained the network with a deep supervision [11] component, which allows better supervision and improves accuracy.

2.2 Dense Criss-Cross Attention (DCCA) Block

A key component in DC^{2}U-Net is the DCCA block, which is designed based on an efficient criss-cross attention mechanism [9] and dense connections [8]. The input peak map is subjected to four consecutive downsampling operations in the encoder branch to obtain the high-level feature map F, and then F is sent to the first criss-cross attention component to generate a new feature map $F^{'}$.

Specifically, F is divided into Q, K, and V three branches, which are all obtained by 1×1 convolutional network dimensionality reduction, where Q, $K \in \mathbb{R}^{C^{'} \times W \times H}$ with $C^{'}$ denoting the number of channels. The attention map $A \in \mathbb{R}^{(H+W-1) \times (W \times H)}$ is calculated using Q and K through affinity operation and is defined as follows

$$d_{i,u} = Q_u \Omega_{i,u}^{\mathrm{T}}, \tag{1}$$

where $Q_u \in \mathbb{R}^{C^{'}}$ is the value of the u position in the feature map Q, and $\Omega_{i,u} \in \mathbb{R}^{C^{'}}$ is the i-th element in Ω_u with $\Omega_u \in \mathbb{R}^{(H+W-1) \times C^{'}}$ denoting the set of elements at the position u on K. It is worth noting that the dimension of Ω_u is significantly reduced in comparison with traditional non-local based self-attention mechanism [20] due to the use of a sparse attention map, which reduces the number of weights from $W \times H$ to $H + W - 1$ [9]. After computing all $d_{i,u}$, we have $D = \{d_{i,u}\}$. Finally, a softmax operation is performed on D in the channel dimension to calculate attention map A.

Another branch V goes through a 1×1 convolutional layer to obtain the adaptive features of $V \in \mathbb{R}^{C \times W \times H}$. We also define $V_u \in \mathbb{R}^{C}$ and $\Phi_u \in \mathbb{R}^{(H+W-1) \times C}$, where Φ_u is the set of feature vectors at the position u on V. Contextual information is learned by aggregation operations defined below

$$F_u^{\prime} = \sum_{i=0}^{H+W-1} A_{i,u} \Phi_{i,u}, \tag{2}$$

where $F_u^{\prime}$ is a feature vector in $F^{\prime} \in \mathbb{R}^{C \times W \times H}$ at position u, and $A_{i,u}$ is a scalar value at channel i and position u in attention map A.

Furthermore, we introduce dense connections [8] into the criss-cross attention component to improve the training efficiency of the network and prevent the gradient from disappearing. Finally, as shown in Fig. 1, the output of the DCCA block, F_{Out} is defined as

$$F_{\mathrm{Out}} = F + F^{'} + F^{'''}, \tag{3}$$

where $F^{'''}$ is the output of the second criss-cross attention component.

2.3 Deeply Supervised Loss Function

A deep supervision strategy [11] is employed to design the loss function. Specifically, as shown in Figure 1, supervision is added to each convolution module of the decoder, which enables better gradient flow and more efficient network training. We use binary cross entropy (BCE) loss as the loss function of our network. Therefore, the overall loss is designed as

$$L = \sum_{k=1}^{5} L_{\text{BCE}}(p_k, g), \tag{4}$$

where g denotes the ground truth, p_k is the k-th prediction output by the network. $L_{\text{BCE}}(p_k, g)$ is defined as

$$L_{\text{BCE}}(p_k, g) = -\frac{1}{N}\sum_{i=0}^{N}(g[i]log(p_k[i]) + (1 - g[i])log(1 - p_k[i])], \tag{5}$$

where N represents the number of classes.

3 Experiments

3.1 Dataset and Implementation Details

We evaluate our DC^{2}U-Net with the dataset in [22], which contains 105 subjects from the Human Connectome Project (HCP) and reference segmentation of 72 major WM tracts per subject. We follow the same dataset partitioning, data augmentation, and experiment settings as in [22]. The data augmentation includes

- Elastic deformation with alpha and sigma$(\alpha, \sigma) \sim (U[90, 120], U[9, 11])$. A displacement vector is sampled for each voxel d $\sim U[-1, 1]$, which is then smoothed by a Gaussian filter with standard deviation σ and finally scaled by α.
- Rotation by angle $\varphi_x \sim U$[-π/4,π/4], $\varphi_y \sim U$[-π/4,π/4], $\varphi_z \sim U$[-π/4,π/4]
- Resampling (to simulate lower image resolution) with factor $\lambda \sim U[0.5, 1]$
- Gaussian noise with mean and variance $(\mu, \sigma) \sim (0, U[0, 0.05])$
- Displacement by (Δ x, Δ y)$\sim (U[-10, 10], U[-10, 10])$
- Zooming by a factor $\lambda \sim U[0.9, 1.5]$

We use three types of inputs: (1) multi-shell multi-tissue CSD using all gradient directions, (2) standard CSD using only b = 1000 s/mm^2 gradient directions, (3) standard CSD using only 12 gradient directions at b = 1000 s/mm^2. Training samples are randomly sampled from these three types of peak maps and the axial, coronal, and sagittal directions of each peak with the size of 144×144 to fit the input of the network. We implement the proposed method using PyTorch [17] on an Nvidia Tesla V100 GPU with 32 GB of memory.

We compare our DC^{2}U-Net with two cutting-edge models. One is recobundles [6], a widely-adopted traditional tract segmentation model. We follow the default parameters of recobundles [6] and use an anatomically-constrained particle filtering probabilistic tractography algorithm [7] to generate whole brain tracts. Another one is TractSeg [22], a powerful deep learning model for tract segmentation. The experimental results of TractSeg are generated using the pretrained model provided in[1].

3.2 Results

The evaluation is performed using the dice score [19], which is one of the widely used metrics in the field of medical image segmentation. We select one testing subject, whose dice score is closest to the average dice score of the entire dataset (ID: 623844), and show the qualitative and quantitative results of the subject in Fig. 2. In the experiment, we also chose the fiber bundles with different reconstruction difficulty levels proposed in the study [15] to represent the segmentation results of the whole fiber list. The degree of difficulty from hard to easy is as follows: anterior commissure (CA), corticospinal tract (CST), and inferior occipito-frontal fascicle (IFO). CA is a fiber bundle that is most difficult to reconstruct among these three bundles since it is relatively slender, making it hard to characterize well with dMRI that suffers from low resolution [3] and heavy noise [2,4]. As shown in Fig. 2, the final dice score of RecoBundles [6] is relatively low due to over-segmentation on CA and IFO_right. Our proposed method has a more significant improvement for small fiber bundles that are difficult to segment, and outperforms comparing segmentation methods for grooves and contours of fibers. Table 1 shows the average dice coefficients between our proposed method and the competing methods on the test dataset with three major comparisons of fiber bundles. Statistically significant paired Student's t-test were also performed for the proposed method and the reference method. Compared with the competing method, the average dice coefficients in the three major fibers are improved, indicating improved performance. These results together demonstrate that our method can significantly improve the performance of WM tract segmentation.

3.3 Ablation Analysis of DC^{2}U-Net

To verify the effectiveness of different components of the proposed method, we perform an extensive ablation study. Similar to Sect. 3.2, we use 21 subjects from the test set for evaluation and randomly select one of them to present the visualization results. CA, FX_left, and FX_right are selected as bounds of interest as they are tiny bundles that are more difficult to segment than other bundles. We use the criss-cross attention block proposed by [9] as a reference for our comparison. As shown in Table 2, the results of the combination of backbone and different blocks are provided. Under the three types of fiber bundles that are difficult to segment, the combination of backbone and DCCA block with deep supervision strategy (DS) is more effective than other blocks. A higher dice score

[1] https://zenodo.org/record/3518348/files/best_weights_ep220.npz.

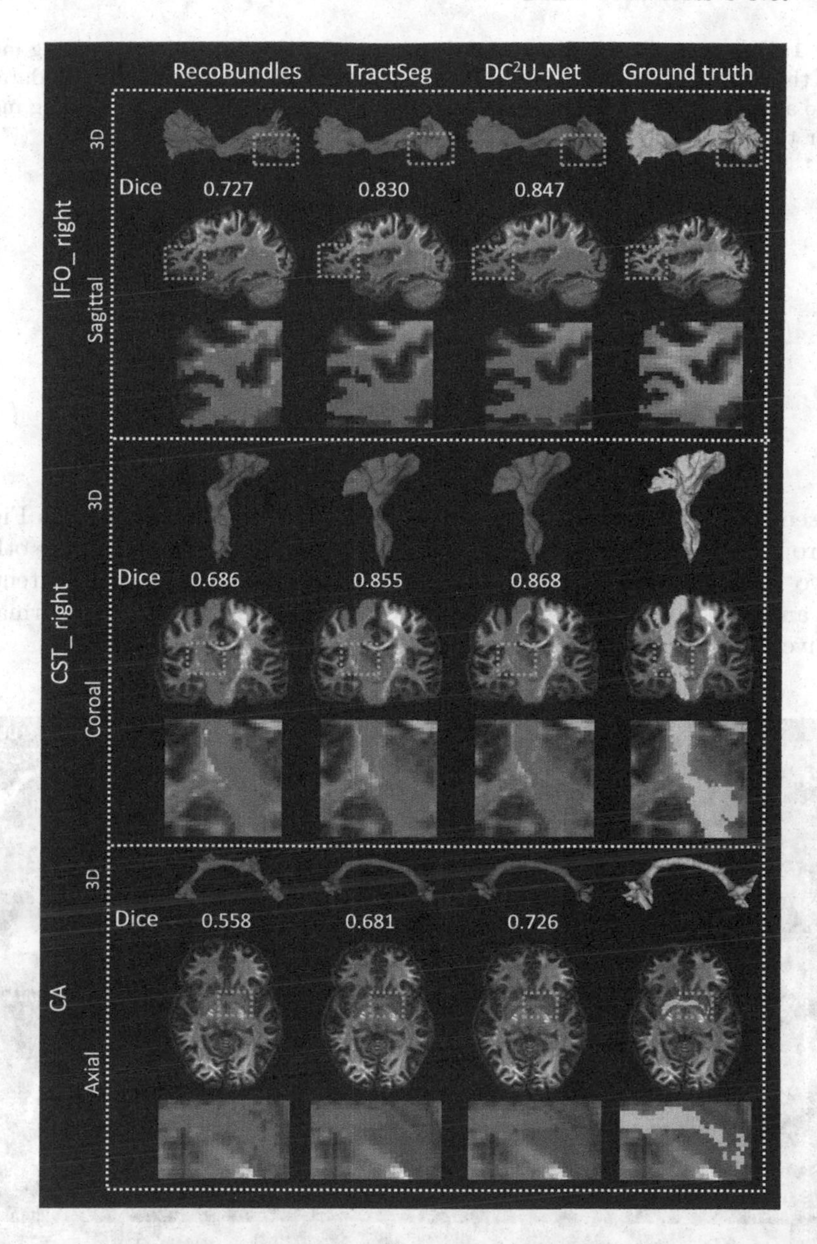

Fig. 2. Qualitative analysis of our method and other methods in comparison. Segmentation results (red) and ground truth (green) are presented in 3D with different views for tract bundles, including the right corticospinal tract (CST), anterior commissure (CA), and right suboccipital frontal (IFO), respectively. The quantitative results of the corresponding fiber bundles are also provided in Table 1 (Color figure online).

is obtained, which is nearly 1.4 % higher than the backbone we used. FX_left and FX_right are relatively symmetrical in the brain, so we only use FX_left to

Table 1. The mean dice coefficients of the proposed method and the competing methods in the three fiber bundles (i.e., CA, CST_right, and IFO_right) on the test dataset. We use a statistically significant t-test method as our evaluation and competing methods for the significance of the difference. The best results are marked in bold. ($^{**}p <$ 0.01 , $^{***}p < 0.001$).

Method	Metric	CA	CST_right	IFO_right	Mean
RecoBundles [6]	Dice	0.539	0.672	0.714	0.654
	p	$***$	$***$	$***$	$***$
TractSeg [22]	Dice	0.682	0.846	0.810	0.829
	p	$***$	**	$***$	$***$
Proposed	Dice	**0.693**	**0.855**	**0.827**	**0.843**

represent this set of visualization results. From the visualization results Fig. 3, our proposed DCCA block with DS is closer to the ground truth than the others on two representative fiber bundles, demonstrating that (i) our self-attention block and deep supervised loss can improve WM tract segmentation performance effectively.

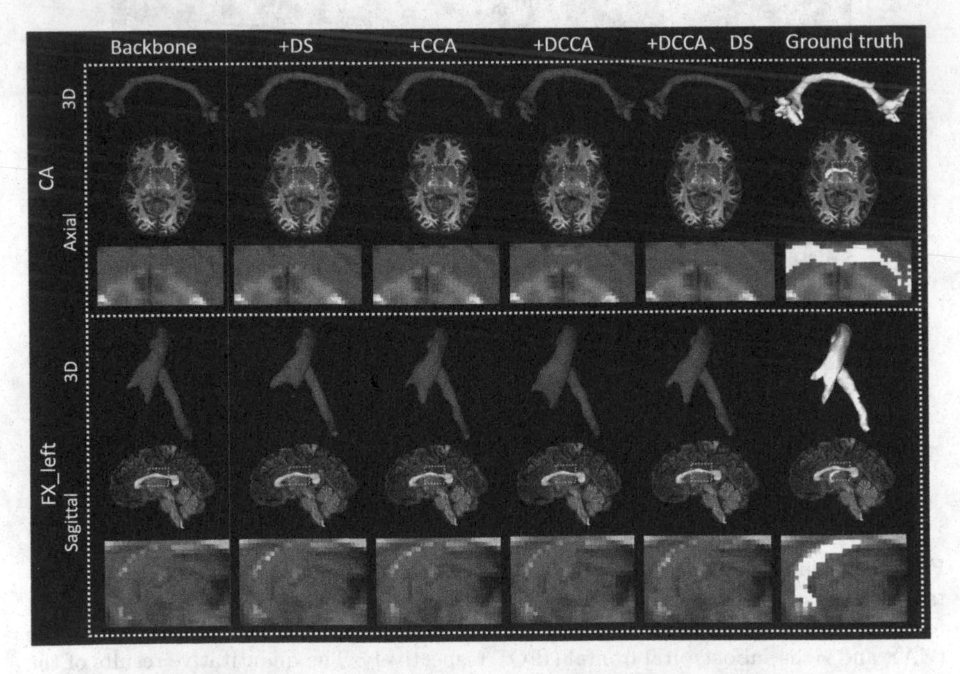

Fig. 3. Visualization results of ablation experiments for the proposed DC^{2}U-Net. Red represents the result of segmenting the fibers using our model and its ablated versions, while green represents the ground truth fiber bundle. (Color figure online)

Table 2. Quantitative results for the ablation study of our DC^{2}U-Net. DS: deep supervision strategy. CCA: criss-cross attention. DCCA: dense criss-cross attention. CA, FX_left, and FX_right represent three different fiber bundles. Mean: Mean dice of the 72 major fiber bundles for each subject in the test set.

Backbone	DS	CCA	DCCA	FX_left	FX_right	CA	Mean
✓				0.718	0.667	0.682	0.829
✓	✓			0.738	0.679	0.685	0.834
✓		✓		0.737	0.682	0.683	0.834
✓			✓	0.743	0.687	0.687	0.839
✓	✓		✓	**0.751**	**0.691**	**0.693**	**0.843**

4 Conclusion

In this paper, we have proposed a novel model, DC^{2}U-Net, for WM fiber tract segmentation. We employ the densely connected self-attention mechanism to exploit the long-range relationships for improving the accuracy with less computational cost. Qualitative and quantitative evaluation results on HCP data demonstrate that our model outperforms cutting-edge methods. Further ablation study verifies the effectiveness of densely connected criss-cross attention and deep supervision strategy.

References

1. Chandio, B.Q., et al.: Bundle analytics, a computational framework for investigating the shapes and profiles of brain pathways across populations. Sci. Rep. **10**(1), 1–18 (2020)
2. Chen, G., Dong, B., Zhang, Y., Lin, W., Shen, D., Yap, P.T.: Denoising of infant diffusion MRI data via graph framelet matching in x-q space. IEEE Trans. Med. Imaging **38**(12), 2838–2848 (2019)
3. Chen, G., Dong, B., Zhang, Y., Lin, W., Shen, D., Yap, P.T.: XQ-SR: joint x-q space super-resolution with application to infant diffusion MRI. Med. Image Anal. **57**, 44–55 (2019)
4. Chen, G., Wu, Y., Shen, D., Yap, P.T.: Noise reduction in diffusion MRI using non-local self-similar information in joint x-q space. Med. Image Anal. **53**, 79–94 (2019)
5. Essen, D.C.V., Smith, S.M., Barch, D.M., Behrens, T.E., Yacoub, E., Ugurbil, K.: The WU-Minn human connectome project: an overview. Neuroimage **80**, 62–79 (2013)
6. Garyfallidis, E., et al.: Recognition of white matter bundles using local and global streamline-based registration and clustering. Neuroimage **170**, 283–295 (2018)
7. Girard, G., Whittingstall, K., Deriche, R., Descoteaux, M.: Towards quantitative connectivity analysis: reducing tractography biases. Neuroimage **98**, 266–278 (2014)

8. Huang, G., Liu, Z., Van Der Maaten, L., Weinberger, K.Q.: Densely connected convolutional networks. In: Proceedings of the IEEE Conference on Computer Vision and Pattern Recognition, pp. 4700–4708 (2017)
9. Huang, Z., Wang, X., Huang, L., Huang, C., Wei, Y., Liu, W.: CCNet: criss-cross attention for semantic segmentation. In: Proceedings of the IEEE International Conference on Computer Vision, pp. 603–612 (2019)
10. Jeurissen, B., Tournier, J.D., Dhollander, T., Connelly, A., Sijbers, J.: Multi-tissue constrained spherical deconvolution for improved analysis of multi-shell diffusion MRI data. Neuroimage **103**, 411–426 (2014)
11. Lee, C.Y., Xie, S., Gallagher, P., Zhang, Z., Tu, Z.: Deeply-supervised nets. In: Artificial Intelligence and Statistics, pp. 562–570. PMLR (2015)
12. Li, B., et al.: Neuro4neuro: a neural network approach for neural tract segmentation using large-scale population-based diffusion imaging. Neuroimage **218**, 116993 (2020)
13. Liu, F., et al.: DeepBundle: fiber bundle parcellation with graph convolution neural networks. In: Zhang, D., Zhou, L., Jie, B., Liu, M. (eds.) GLMI 2019. LNCS, vol. 11849, pp. 88–95. Springer, Cham (2019). https://doi.org/10.1007/978-3-030-35817-4_11
14. Lu, Q., Li, Y., Ye, C.: Volumetric white matter tract segmentation with nested self-supervised learning using sequential pretext tasks. Med. Image Anal. **72**, 102094 (2021)
15. Maier-Hein, K.H., et al.: The challenge of mapping the human connectome based on diffusion tractography. Nat. Commun. **8**(1), 1–13 (2017)
16. Mou, L., et al.: Cs2-net: deep learning segmentation of curvilinear structures in medical imaging. Med. Image Anal. **67**, 101874 (2021)
17. Paszke, A., et al.: PyTorch: an imperative style, high-performance deep learning library. In: Advances in Neural Information Processing Systems, vol. 32 (2019)
18. Ronneberger, O., Fischer, P., Brox, T.: U-net: convolutional networks for biomedical image segmentation. In: Navab, N., Hornegger, J., Wells, W.M., Frangi, A.F. (eds.) MICCAI 2015. LNCS, vol. 9351, pp. 234–241. Springer, Cham (2015). https://doi.org/10.1007/978-3-319-24574-4_28
19. Taha, A.A., Hanbury, A.: Metrics for evaluating 3D medical image segmentation: analysis, selection, and tool. BMC Med. Imaging **15**(1), 1–28 (2015)
20. Wang, X., Girshick, R., Gupta, A., He, K.: Non-local neural networks. In: Proceedings of the IEEE Conference on Computer Vision and Pattern Recognition, pp. 7794–7803 (2018)
21. Wassermann, D., et al.: The white matter query language: a novel approach for describing human white matter anatomy. Brain Struct. Funct. **221**(9), 4705–4721 (2016)
22. Wasserthal, J., Neher, P., Maier-Hein, K.H.: TractSeg-fast and accurate white matter tract segmentation. Neuroimage **183**, 239–253 (2018)
23. Zarkali, A., McColgan, P., Leyland, L.A., Lees, A.J., Rees, G., Weil, R.S.: Fiber-specific white matter reductions in Parkinson hallucinations and visual dysfunction. Neurology **94**(14), e1525–e1538 (2020)

Clustering in Tractography Using Autoencoders (CINTA)

Jon Haitz Legarreta[1,2(✉)], Laurent Petit[3], Pierre-Marc Jodoin[2,4], and Maxime Descoteaux[1,4]

[1] Sherbrooke Connectivity Imaging Laboratory (SCIL), Department of Computer Science, Université de Sherbrooke, Sherbrooke, Québec, Canada
{jh.legarreta,m.descoteaux}@usherbooke.ca

[2] Videos & Images Theory and Analytics Laboratory (VITAL), Department of Computer Science, Université de Sherbrooke, Sherbrooke, Québec, Canada
pierre-marc.jodoin@usherbooke.ca

[3] Groupe d'Imagerie Neurofonctionnelle (GIN), Univ. Bordeaux, CNRS, CEA, IMN, UMR 5293, France
laurent.petit@u-bordeaux.fr

[4] Imeka Solutions inc., Sherbrooke, Québec, Canada

Abstract. Clustering tractography streamlines is an important step to characterize the brain white matter structural connectivity. Numerous methods have been proposed to group whole-brain tractography streamlines into anatomically coherent bundles. However, the time complexity, or the initial streamline sorting in conventional methods, or still, using supervised deep learning models, may limit the results and/or restrict the versatility of the methods. In this work, we propose an autoencoder-based method for clustering tractography streamlines. CINTA, *Clustering in Tractography using Autoencoders*, is trained on unlabelled data, uses a single autoencoder model, and does not require any distance thresholding parameter. It obtains excellent classification scores on synthetic datasets, achieving a 0.97 F1-score on the clinical-style, realistic ISMRM 2015 Tractography Challenge dataset. Similarly, CINTA obtains anatomically reliable results on *in vivo* human brain tractography data. CINTA offers a time-efficient bundling framework, as its running time is linear with the streamline count.

Keywords: Representation Learning · Autoencoder · diffusion MRI · Tractography · Clustering

1 Introduction

White matter (WM) brain fiber parcellation, also named *bundling*, or segmentation –especially when providing a voxel-based output–, or "virtual dissection" when being done semi-automatically or with some manual intervention,

P.-M. Jodoin and M. Descoteaux—Co-senior authors. These authors contributed equally.

S. Cetin-Karayumak et al. (Eds.): CDMRI 2022, LNCS 13722, pp. 125–136, 2022.
https://doi.org/10.1007/978-3-031-21206-2_11

encompasses methods that aim to classify and group together fiber entities, i.e. streamlines. Bundling is an essential processing step in tractography pipelines allowing to identify the tracks of interest across different brain regions. The large number of streamlines contained in an average tractogram calls for automated procedures. Streamline classification for bundling purposes is most commonly performed using either of two criteria [6]: (i) the streamline similarity (defined according to some distance measure); and/or (ii) the regions of interest (ROI) streamlines traverse or which (gray matter) brain regions their endpoints connect. Despite being a seemingly simple geometrical entity, adequately characterizing streamlines is still a challenge. Although several distance measures (such as the closest point distance, the Hausdorff distance, the Mahalanobis distance, or the Minimum average Direct and Flip distance (MDF), among others) have been proposed in literature [6,17], streamline-space point-wise distance computation and full pair-wise comparisons are computationally expensive, and might not capture other relevant features. Clustering can be performed in the streamline native space, or some other representation space (e.g. [17,23,27]), and some methods provide a volumetric result of streamline groups (bundles) (e.g. [13,14,22]).

We propose to extend the autoencoder-based latent space nearest neighbor tractography framework proposed in [12] to cluster streamlines into bundles. We show that the proposed autoencoder-based method is successful at bundling streamlines on synthetic and clinical-style realistic phantom and *in vivo* human brain data. The method (i) does not require to be trained on labelled data, (ii) uses a single model, trained only once, to classify streamlines, and (iii) does not require any distance thresholding parameter to generate the clusters.

1.1 Related Work

Automatic bundle identification of deep white matter pathways has been performed using a variety of methods: (i) anatomical filtering; (ii) clustering; (iii) atlas-based; (iv) graph-based; (v) dictionary learning;(vi) segmentation-based; and, more recently, (vii) deep learning-based methods [23]. Automatic anatomical filtering methods (e.g. [26]), including query languages [21], often offer limited quality results due to the variability of the streamline locations across subjects, and are highly sensitive to the streamlines' waypoints (e.g. streamlines that are a few voxels short of reaching the gray matter, or apart from each other at a few locations might be discarded or classified into different groups).

Clustering methods [2,9,15,17,19] use a given streamline similarity distance definition. These approaches may include some form of hierarchical approach to progressively improve the results (e.g. [9]). Several methods have used unsupervised machine learning strategies, such as Expectation-Maximization (EM) [15] or k-means [9]. Similarly, the use of streamline feature descriptors that aim to capture and summarize the relevant information for the classification, along with the use of some form of embedding space where the clustering takes place, have also been proposed [2,17,19]. Some of these methods (e.g. [17]) require computing pair-wise streamline distances, which has a complexity of $\mathcal{O}(N^2)$.

Atlas-based methods such as the ones proposed in [6,25] rely on the anatomical priors provided by the atlas to assign streamlines to a given bundle. They use bundle or cluster "models" to recognize streamlines in the target tractogram according to a given threshold with respect to the streamline- or feature-space centroids. Some of these methods, such as [6], might yield a variable number of clusters across subjects, or differing results depending on the initial sorting of the streamlines in the tractogram.

Graph-based strategies [18,20] consider the clustering task as a graph partitioning problem that seeks to cluster the nodes based on a similarity measure. Dictionary learning methods [23], in turn, generally assume that a dictionary that contains a representative signature for each bundle can be computed (or learned), and posit the task of finding the class a streamline belongs to as an optimization problem that seeks to find the coefficients that fit a given bundle representation for each streamline.

Lately, deep learning-based methods have also been applied to the bundling task, and have compared favorably over the mentioned conventional methods within the studied contexts. Several authors [11,27] have used recurrent neural networks (RNNs) to solve the clustering problem as a classification problem. Similarly, regular classification convolutional neural networks (CNNs) have been employed [10,19,24] to predict the streamline bundle labels. In [3], authors proposed a Deep Embedded Clustering-based (DEC) framework to provide the cluster assignments. Finally, a number of deep learning-based methods have cast the problem into a segmentation task, yielding bundle-wise voxel masks [13,14,22].

Classification neural networks are trained to reliably provide a prediction on a fixed-length probability vector, and hence do not allow to change the number of target labels (i.e. bundles) without retraining. Tractography segmentation methods, in turn, are inherently binary classification methods: given that the same voxel cannot be assigned to multiple labels (even if multiple streamlines belonging to different bundles may traverse the same voxel), such methods require a separate model to be trained for each bundle.

2 Material and Methods

The same deep autoencoder architecture presented in [12] is used in this work. The chosen autoencoder is a regular convolutional deep neural network, trained to minimize the mean squared-error loss between the input streamlines and their reconstructions at the output of the autoencoder.

We propose to cluster streamlines using a k-NN approach in the latent space learned by autoencoding streamlines. It is essentially assumed that similar data points (streamlines in our case) will be concentrated to neighboring regions in the Euclidean sense in the latent space [1,8]. Thus, given (i) an autoencoder; (ii) a set of streamlines to train the autoencoder; (iii) the anatomical bundle classes of a subset of the preceding streamlines; and (iv) a new tractogram that needs to be split into the same set of available bundles, the proposed method proceeds as follows:

1. Train an autoencoder using raw, unlabelled streamlines, generated by a predetermined tractography algorithm.
2. Select a subset of streamlines whose bundle class is known so that they can be used as the reference set to bundle new streamlines. Project such streamlines to the latent space.
3. Project to the latent space the streamlines in a new, to-be-bundled tractogram.
4. Apply a k-NN method using the readily available labelled (reference) streamlines to determine the bundle class of the new streamlines.

We have dubbed the above method CINTA, *Clustering in Tractography using Autoencoders*. The method requires all streamline data to dwell in a common or standard reference space (such as the MNI space).

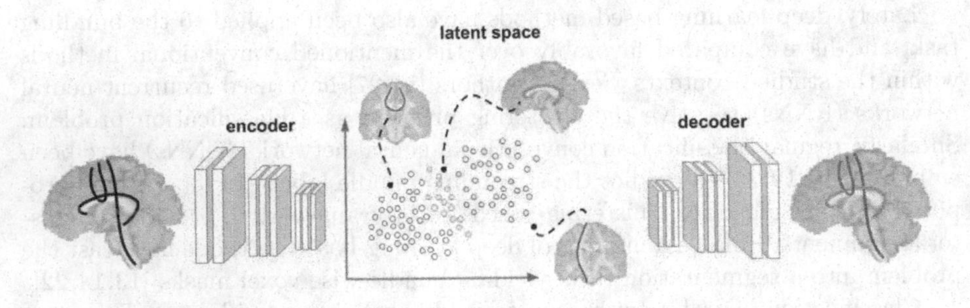

Fig. 1. Conceptual illustration of CINTA (*Clustering in Tractography using Autoencoders*). The streamlines that belong to the same bundle are naturally clustered together in the latent space of a trained autoencoder. A k-NN method is applied to assign the bundle label to such streamlines.

3 Experiments

CINTA's performance is quantitatively measured on the (i) "Fiber Cup" synthetic tractography dataset [4,5], and the (ii) clinical-style realistic ISMRM 2015 Tractography Challenge dataset [16]. A subject from the Human Connectome Project (HCP) dataset [7] was used to qualitatively demonstrate CINTA's bundling ability on *in vivo* human brain tractography data. Local probabilistic ("Fiber Cup"; ISMRM 2015 Tractography Challenge) and global tracking (HCP) were employed to reconstruct streamlines. The ground truth WM parcellations were obtained according to the data preparation procedure described in [12]. Streamlines had their head-to-tail orientations flipped according to a reference, and were resampled to 256 points prior to training the autoencoder. The k parameter for the k-NN clustering method was chosen experimentally from the set 3,5: it was fixed to a value of 5 as it provided a better F1-score on the ISMRM 2015 Tractography Challenge dataset (an identical performance was registered for both values on the "Fiber Cup" dataset). RecoBundles [6] was used as the baseline method (using the synthetic bundle models available in each dataset).

The following results are reported:

- *Accuracy*: proportion of correct predictions (true positives and true negatives) over the total number of streamlines.
- *Sensitivity* (recall): proportion of relevant instances that are predicted as positives (true positives) among all positive streamlines in the data.
- *Precision*: proportion of relevant instances that are predicted as true positives among all retrieved (predicted) positive streamlines.
- *F1-score*: harmonic mean of precision and sensitivity.

For each bundle, the positive instances are those corresponding to the streamlines that are labelled with the given bundle class as determined by the underlying scoring method, the negatives being any other streamline in the whole tractogram.

4 Results

Table 1 shows CINTA's performance for the "Fiber Cup" and ISMRM 2015 Tractography Challenge datasets averaged over all bundles. As the reported measures reveal, the proposed autoencoder-based tractography bundling procedure achieves perfect and close to perfect scores on the respective datasets, and outperforms the RecoBundles baseline consistently. Additionally, as it can be seen in figure 2, the classification performance is highly consistent across bundles on both datasets.

Table 1. Bundling classification scores. Mean and standard deviation values across bundles.

Dataset	Method	Accuracy	Sensitivity	Precision	F1-score
"Fiber Cup"	RecoBundles	0.98 (0.04)	0.99 (0.02)	0.96 (0.11)	0.97 (0.09)
	CINTA	**1.0**	**1.0**	**1.0**	**1.0**
ISMRM 2015	RecoBundles	0.99 (0.01)	**0.99** (0.01)	0.88 (0.15)	0.91 (0.12)
	CINTA	**1.0**	0.97 (0.04)	**0.97** (0.04)	**0.97** (0.04)

Figures 3 and 4 show the bundles as classified with the proposed method. As expected from the scores in table 1, the latent space-based bundling predictions closely follow the anatomically coherent streamline-space bundle partitions. Furthermore, following from the reconstruction difficulty analysis on the ISMRM 2015 Tractography Challenge dataset [16], which revealed 18 hard or very hard bundles, results indicate that CINTA reliably identifies hard-to-track bundles in the data (e.g. left CST and fornix; see (e) and (f) subplots in figure 4).

Figure 5 shows the bundling results on the HCP data subject. As it detaches from the figure, CINTA successfully clusters streamlines into the corresponding anatomically meaningful bundles.

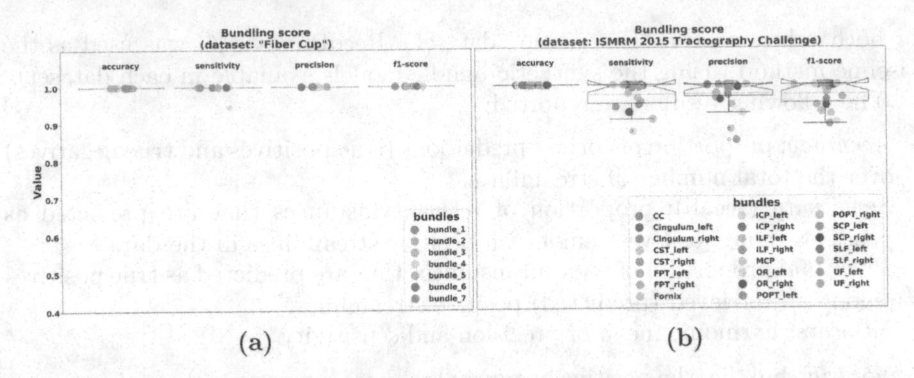

(a) (b)

Fig. 2. CINTA's classification performance bundle-wise breakup: (a) "Fiber Cup" dataset; and (b) ISMRM 2015 Tractography Challenge dataset.

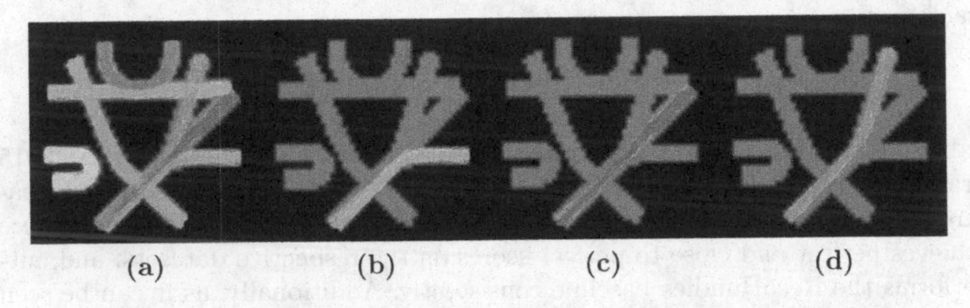

(a) (b) (c) (d)

Fig. 3. Autoencoder-based bundling on the "Fiber Cup" dataset: (a) all bundles; (b) bundle 5; (c) bundle 6; and (d) bundle 7 (following the numbering in [4]).

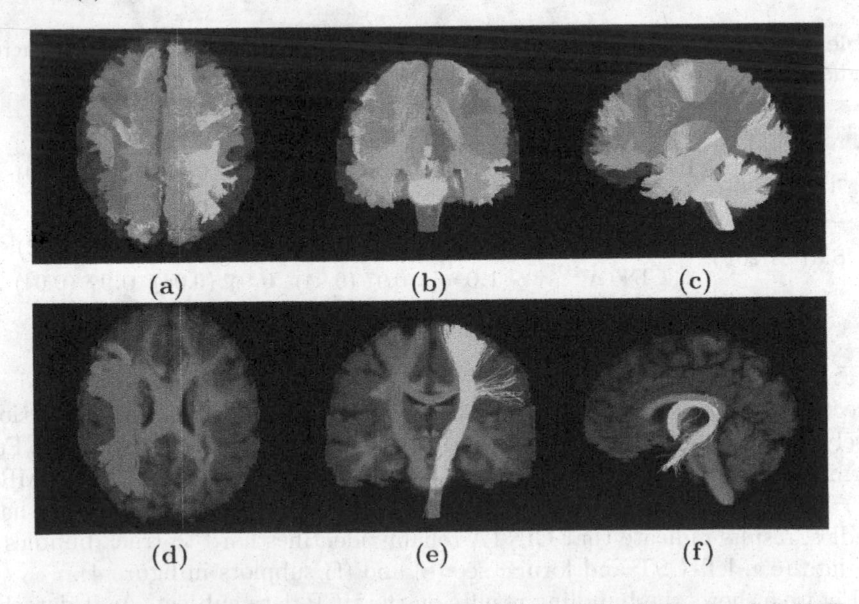

(a) (b) (c)

(d) (e) (f)

Fig. 4. Autoencoder-based bundling on the ISMRM 2015 Tractography Challenge dataset: (a, b, c) all bundles (axial superior, coronal anterior, sagittal left views, respectively); (d) left SLF (axial superior view); (e) left CST (coronal anterior view); and (f) Fornix (sagittal left view) (see [16] for the bundle acronyms and names).

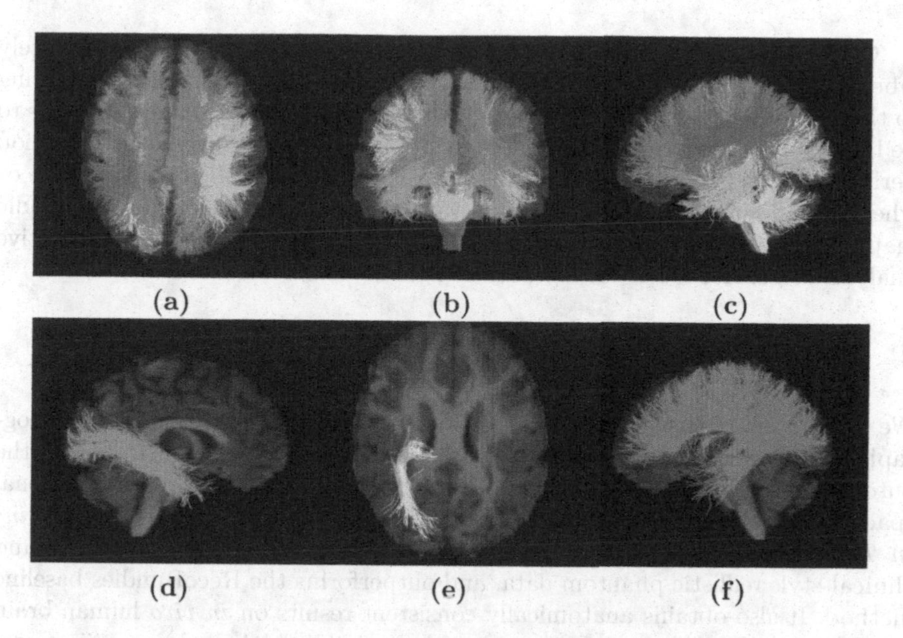

Fig. 5. Autoencoder-based bundling on the HCP dataset: (a, b, c) all bundles (axial superior, coronal anterior, sagittal left views, respectively); (d) right ILF (sagittal right view); (e) left OR (axial superior view); and (f) CC (sagittal left view).

5 Discussion

The results in section 4 show that the latent space learned by the proposed autoencoder provides a low-dimensional representational space where *similar* streamlines are clustered close to each other. Thus, streamlines can be appropriately classified into anatomically coherent bundles in such a space.

Our clustering approach only requires a single parameter to be fixed (the neighborhood value k), and it is experimentally verified that its value does not influence significantly the results. Its worst case computational time performance is linear ($\mathcal{O}(Nd) \approx \mathcal{O}(N)$, where N is the number of data points and d the number of features, assuming $N \gg d$) (see section A.2 for an experimental demonstration). The complexity is thus dominated by the number of samples. Our clustering framework uses a single model to classify all streamlines at once. Additionally, CINTA can accommodate a variable number of bundles: the autoencoder does not need to be retrained if the number of bundles to be identified changes.

The proposed procedure does not incur notable misclassification errors: it is verified that when a streamline is assigned to the wrong bundle, such streamlines are anatomically close to the wrong class (e.g. left CST streamlines being classified as left FPT streamlines; see section A.1 for an example). This constitutes an indirect evidence of the fact that the latent space of our autoencoder appropriately encodes the necessary anatomical information about the input streamlines.

CINTA requires a subset of the training streamlines to be appropriately labelled so that streamlines in any new tractogram can be classified according to their nearest neighbors in such set. Such a set of labelled streamlines needs to be built only once (for a target bundle mapping). Investigating the classification performance dependency on the number of available labelled streamlines, or whether and how such a value may be variable across bundles or target bundle mappings, is left for future work. Similarly, a multi-subject dataset comparative analysis of CINTA is left for a separate piece of work.

6 Conclusion

We present an extension to an autoencoder-based framework to cluster tractography streamlines into anatomically consistent bundles. We demonstrate that the autoencoder-based tractography latent space offers a versatile representational space to classify streamlines in a straightforward fashion. CINTA (*Clustering in Tractography using Autoencoders*), obtains excellent scores in synthetic and clinical-style realistic phantom data, and outperforms the RecoBundles baseline method. It also obtains anatomically consistent results on *in vivo* human brain data. The method (i) does not require to be trained on labelled data, (ii) uses a single model, trained only once, to classify streamlines, and (iii) does not require any distance thresholding parameter to generate the clusters.

Conflict of Interest. Pierre-Marc Jodoin and Maxime Descoteaux report a relationship with Imeka Solutions inc. that includes board membership and employment. Jon Haitz Legarreta, Pierre-Marc Jodoin and Maxime Descoteaux have patent #17/337,413 pending to Imeka Solutions inc.

Acknowledgments. This work has been partially supported by the Centre d'Imagerie Médicale de l'Université de Sherbrooke (CIMUS); the Axe d'Imagerie Médicale (AIM) of the Centre de Recherche du CHUS (CRCHUS); and the Réseau de Bio-Imagerie du Québec (RBIQ)/Quebec Bio-imaging Network (QBIN) (FRSQ - Réseaux de recherche thématiques File: 35450). This research was enabled in part by support provided by the Digital Research Alliance of Canada Advanced Research Computing service (alliancecan.ca) and Calcul Québec (calculquebec.ca). We also thank the research chair in Neuroinformatics of the Université de Sherbrooke.

A Appendix

A.1 Misclassified Streamlines

Figure 6 shows the split of the ISMRM 2015 Tractography Challenge right FPT bundle as classified by the autoencoder-based bundling procedure. As it detaches from the figure, the misclassified streamlines belong to bundles (right CST and right POPT) that are closely related to it in anatomical and/or spatial terms. This reinforces the assumption that streamlines that are close to each other in anatomical space are also located in neighboring regions in the latent space learned by the CINTA autoencoder. Hence, CINTA provides an anatomically reliable ground for bundling purposes with minimal disagreement.

A.2 Time Computational Requirements

To demonstrate CINTA's computational time performance, six (6) tractograms containing 20 000, 40 000, 100 000, 200 000, 600 000, 1 000 000 streamlines were generated on the ISMRM 2015 Tractography Challenge dataset using local probabilistic tracking. Implausible streamlines were filtered following the method proposed in [12]. The time required to bundle each resulting tractogram was measured for three (3) runs, and the mean and standard deviation values computed. Only the time required for bundling was measured, excluding I/O operation time. Time tests were performed on a conventional desktop machine (Intel(R) Xeon(R) W-2133 CPU @3.60 GHz 6 core processor; 16 G RAM; NVIDIA GeForce GTX 1080 Ti 12 G graphics card). As shown in figure 7, CINTA requires a linear time to bundle streamlines. Similarly, its time demands are comparable to other competitive deep learning-based methods reported in literature [3], requiring slightly less than 200 s to bundle almost 600 000 streamlines.

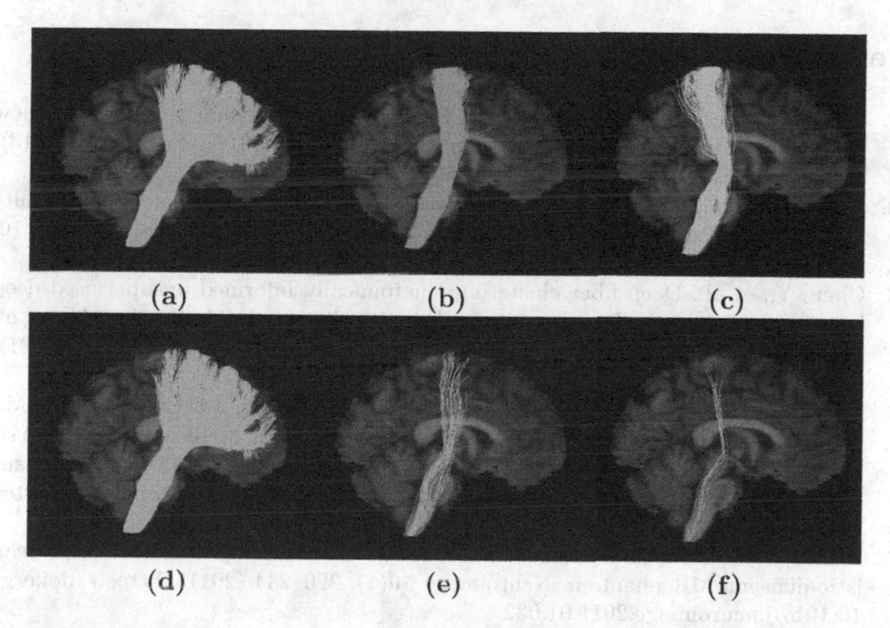

Fig. 6. Right FPT bundle of the ISMRM 2015 Tractography Challenge dataset as labelled by the autoencoder-based bundling procedure: (a) right FPT; (b) right CST in the reference set; (c) right POPT in the reference set; (d) right FPT true positives; (e) false positive right FPT streamlines belonging to the right CST; (f) false positive right FPT streamlines belonging to the right POPT. All sagittal right views.

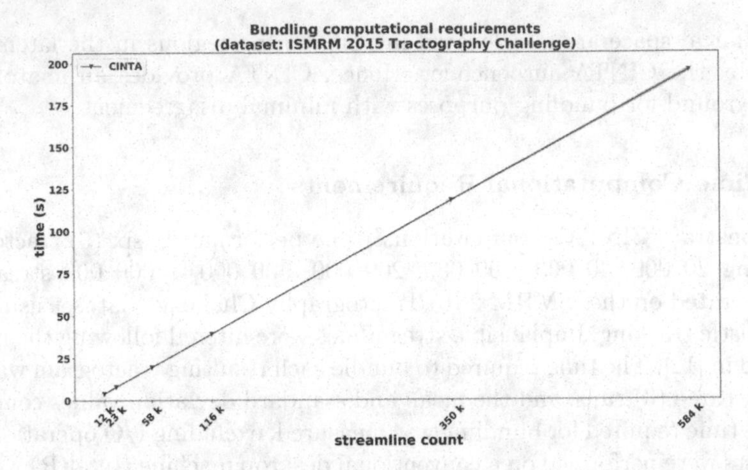

Fig. 7. Computational time performance for bundling different ISMRM 2015 Tractography Challenge dataset tractogram sizes with CINTA. Due to the vertical scale and reduced standard deviation values, the latter are hardly noticeable around the mean value. Streamline counts are expressed with SI prefixes and engineering notation. Horizontal axis labels correspond to filtered tractogram streamline counts.

References

1. Bengio, Y., Courville, A., Vincent, P.: Representation learning: a review and new perspectives. IEEE Trans. Pattern Anal. Mach. Intell. **35**(8), 1798–1828 (2013). https://doi.org/10.1109/TPAMI.2013.50
2. Bertò, G., et al.: Classifyber, a robust streamline-based linear classifier for white matter bundle segmentation. Neuroimage **224**, 117402 (2021). https://doi.org/10.1016/j.neuroimage.2020.117402
3. Chen, Y., et al.: Deep fiber clustering: anatomically informed unsupervised deep learning for fast and effective white matter parcellation. In: de Bruijne, M., et al. (eds.) MICCAI 2021. LNCS, vol. 12907, pp. 497–507. Springer, Cham (2021). https://doi.org/10.1007/978-3-030-87234-2_47
4. Côté, M.A., Girard, G., Boré, A., Garyfallidis, E., Houde, J.C., Descoteaux, M.: Tractometer: Towards validation of tractography pipelines. Medical Image Analysis 17(7), 844–857 (2013). https://doi.org/10.1016/j.media.2013.03.009, special Issue on the 2012 Conference on Medical Image Computing and Computer Assisted Intervention
5. Fillard, P., et al.: Quantitative evaluation of 10 tractography algorithms on a realistic diffusion MR phantom. Neuroimage **56**(1), 220–234 (2011). https://doi.org/10.1016/j.neuroimage.2011.01.032
6. Garyfallidis, E., et al.: Recognition of white matter bundles using local and global streamline-based registration and clustering. Neuroimage **170**, 283–295 (2018). https://doi.org/10.1016/j.neuroimage.2017.07.015. Segmenting the Brain
7. Glasser, M.F., et al.: The human connectome project's neuroimaging approach. Nat. Neurosci. **19**(9), 1175–1187 (2016). https://doi.org/10.1038/nature18933
8. Goodfellow, I., Bengio, Y., Courville, A.: Deep Learning, Adaptive Computation and Machine Learning. MIT Press, Cambridge (2016)

9. Guevara, P., et al.: Robust clustering of massive tractography datasets. Neuroimage **54**(3), 1975–1993 (2011). https://doi.org/10.1016/j.neuroimage.2010.10.028
10. Gupta, V., Thomopoulos, S.I., Corbin, C.K., Rashid, F.M., Thompson, P.M.: FiberNet 2.0: an automatic neural network based tool for clustering white matter fibers in the brain. In: IEEE 15th International Symposium on Biomedical Imaging (ISBI), pp. 708–711. Washington, DC, USA (04 2018). https://doi.org/10.1109/ISBI.2018.8363672
11. Lam, P.D.N., Belhomme, G., Ferrall, J., Patterson, B., Styner, M., Prieto, J.C.: TRAFIC: Fiber tract classification using deep learning. In: Proceedings of the International Society for Optical Engineering (SPIE). vol. 10574, p. 1057412. The international society for optics and photonics (SPIE) (2018). https://doi.org/10.1117/12.2293931
12. Legarreta, J.H., et al.: Filtering in tractography using autoencoders (FINTA). Medical Image Analysis 72, 102126 (2021). https://doi.org/10.1016/j.media.2021.102126
13. Li, B., et al.: Neuro4Neuro: a neural network approach for neural tract segmentation using large-scale population-based diffusion imaging. Neuroimage (2020). https://doi.org/10.1016/j.neuroimage.2020.116993
14. Liu, W., et al.: Volumetric segmentation of white matter tracts with label embedding. Neuroimage **250**, 118934 (2022). https://doi.org/10.1016/j.neuroimage.2022.118934
15. Maddah, M., Grimson, W.E.L., Warfield, S.K., Wells, W.M.: A unified framework for clustering and quantitative analysis of white matter fiber tracts. Med. Image Anal. **12**(2), 191–202 (2008). https://doi.org/10.1016/j.media.2007.10.003
16. Maier-Hein, K.H., et al.: The challenge of mapping the human connectome based on diffusion tractography. Nat. Commun. **8**(1349), 1–13 (2017). https://doi.org/10.1038/s41467-017-01285-x
17. O'Donnell, L.J., Westin, C.F.: Automatic tractography segmentation using a high-dimensional white matter atlas. IEEE Trans. Med. Imaging **26**(11), 1562–1575 (2007). https://doi.org/10.1109/TMI.2007.906785
18. Siless, V., Chang, K., Fischl, B., Yendiki, A.: AnatomiCuts: hierarchical clustering of tractography streamlines based on anatomical similarity. Neuroimage **166**, 32–45 (2018). https://doi.org/10.1016/j.neuroimage.2017.10.058
19. Ugurlu, D., Firat, Z., Ture, U., Unal, G.: Supervised classification of white matter fibers based on neighborhood fiber orientation distributions using an ensemble of neural networks. In: Bonet-Carne, E., Grussu, F., Ning, L., Sepehrband, F., Tax, C.M.W. (eds.) MICCAI 2019. MV, pp. 143–154. Springer, Cham (2019). https://doi.org/10.1007/978-3-030-05831-9_12
20. Vázquez, A., et al.: FFClust: fast fiber clustering for large tractography datasets for a detailed study of brain connectivity. NeuroImage 220, 117070 (2020). https://doi.org/10.1016/j.neuroimage.2020.117070
21. Wassermann, D., et al.: The white matter query language: a novel approach for describing human white matter anatomy. Brain Structure and Function 221(9), 4705–4721 (2016). https://doi.org/10.1007/s00429-015-1179-4
22. Wasserthal, J., Neher, P.F., Maier-Hein, K.H.: TractSeg - fast and accurate white matter tract segmentation. NeuroImage 183, 239–253 (2018)
23. Wu, Y., Hong, Y., Ahmad, S., Lin, W., Shen, D., Yap, P.-T.: Tract dictionary learning for fast and robust recognition of fiber bundles. In: Martel, A.L., et al. (eds.) MICCAI 2020. LNCS, vol. 12267, pp. 251–259. Springer, Cham (2020). https://doi.org/10.1007/978-3-030-59728-3_25

24. Zhang, F., Cetin Karayumak, S., Hoffmann, N., Rathi, Y., Golby, A.J., O'Donnell, L.J.: Deep white matter analysis (DeepWMA): fast and consistent tractography segmentation. Medical Image Analysis 65, 101761 (2020). https://doi.org/10.1016/j.media.2020.101761
25. Zhang, F., et al.: Whole brain white matter connectivity analysis using machine learning: an application to autism. Neuroimage **172**, 826–837 (2018). https://doi.org/10.1016/j.neuroimage.2017.10.029
26. Zhang, Y., et al.: Atlas-guided tract reconstruction for automated and comprehensive examination of the white matter anatomy. Neuroimage **52**(4), 1289–1301 (2010). https://doi.org/10.1016/j.neuroimage.2010.05.049
27. Zhong, S., Chen, Z., Egan, G.: Auto-encoded latent representations of white matter streamlines. In: 28th Virtual Conference & Exhibition of the International Society for Magnetic Resonance in Medicine (ISMRM). International Society for Magnetic Resonance in Medicine (2020), abstract #0850

Tractometric Coherence of Fiber Bundles in DTI

Rick Sengers[1](✉), Tom Dela Haije[2], Andrea Fuster[1], and Luc Florack[1]

[1] Eindhoven University of Technology, Eindhoven, The Netherlands
{H.J.C.E.Sengers,A.Fuster,L.M.J.Florack}@tue.nl
[2] University of Copenhagen, Copenhagen, Denmark
tom@di.ku.dk

Abstract. Based on a diffusion tensor image (DTI) and a tentative tractogram of a fiber bundle we propose a filtering method for operationally defining and removing outliers using tractometry. To this end we assign to each track a set of K invariants, i.e. scalars invariant under rigid transformations. The cluster of K-tuples of all tracks in a bundle may be pruned using outlier detection methods in $\mathbb{R}^K$, after which back-projection of the remaining K-tuples produces a filtered tractogram with enhanced coherence. This intrinsic pruning method is blind to the relative spatial organization of tracks in a bundle. We consider two types of invariants, one capturing local diffusion properties and one representing differential properties averaged along tracks. Our experiments indicate that our tractometric filtering is complementary to extrinsic methods based on the relative spatial configuration of tracks.

Keywords: Diffusion tensor imaging · Tractography · Tractometry · Tractography filtering · Riemannian geometry

1 Introduction

Tractography aims at reconstructing fiber bundles in the brain from diffusion weighted imaging (DWI), a non-invasive technique for in-vivo imaging of the brain's fibrous structure. For our purpose tractography may refer to any method providing a bundle of tentative tracks, the most prevalent being either streamline methods based on diffusion tensor imaging (DTI) [2,4,28,37], or geodesic methods [11,15,16,20,25,28,30], whether deterministic or probabilistic.

Given two regions of interest (ROI) any of these methods may yield a collection of putative tracts in-between. In streamline methods this is achieved by designating one ROI as the seed region from which tracks are initialised to define integral curves of some a priori preferred diffusion direction, e.g. the main eigendirection of the diffusion tensor in the case of deterministic tractography based on DTI. In this case the second ROI serves as an include-region which ensures only tracks that pass through it are kept in the tractogram. In geodesic tractography tracks are curves of (locally) shortest length in a Riemannian (or more generally, Finslerian) space, i.e. geodesics [1,17,27,29,32]. The associated

S. Cetin-Karayumak et al. (Eds.): CDMRI 2022, LNCS 13722, pp. 137–148, 2022.
https://doi.org/10.1007/978-3-031-21206-2_12

metric is constructed on the basis of the DTI (or DWI) data in such a way that small distances in Riemannian (or Finslerian) sense are tantamount to large mean free paths. In this case both ROIs are necessary for disambiguation due to the second order nature of geodesic tractography. Geodesic completeness (aka the Hopf-Rinow Theorem) ensures that at least one connection between any two given endpoints exists.

In all cases we obtain as a starting point a single-bundle tractogram consisting of, say, N tracks. This tractogram may contain 'incoherent' fibers, i.e. outliers (not necessarily false positives) that significantly deviate from the main bundle in one way or another. Our goal is to obtain a filtered tractogram of $M \leq N$ selected tracks that, in some precise sense, exhibit more coherence. Instead of utilizing solely spatial information, e.g. as in [13,14,19,22,26] or streamline densities [33] or the diffusion signal [10], we aim to do this by means of *tractometry*, the assignment of characteristic scalars to each track. Tractometry has been used for various purposes, such as dimensionality reduction [7], tract-analysis [38] and anomaly detection [6]. Tractometric scalars can also be used to prune a tractogram by eliminating $N-M$ tracks (with M an automatic or manual control parameter) that are deemed deviatory according to some coherence measure. Underlying this is the assumption that microstructural similarities within a particular anatomical (sub)bundle are reflected in a (macroscopic) similarity among the assigned scalars for that (sub)bundle, i.e. assuming no counteracting pathological effects [6,35]. For this reason we stress that we do not consider the method to work globally on a whole brain tractogram, but as a a single-bundle method pruning its streamlines one by one.

2 Theory

In principle any ad hoc set of scalar functionals could be used in the proposed tractometry framework. However, we will formulate a set of criteria, the first of which pertains to *invariance* under rotations. If f is a scalar function of a diffusion tensor D, it is invariant under rotations if $f(D) = f(R^T DR)$ for any rotation matrix R. We will refer to such invariant scalar functions simply as 'invariants'. An example of a non-invariant scalar would be any isolated component of the diffusion tensor D, whereas the trace $\mathrm{Tr}(D)$ *is* an invariant. Global, track-wise invariants may be constructed from point-wise invariants sampled along a track by taking their average, median, minimum, maximum, or other integral measures. In the rest of the paper we will use the mean for the sake of definiteness.

Secondly, we require an admissible set of invariants to satisfy well-defined *non-redundancy* and *completeness* conditions. These conditions capture the notion of the set containing 'precisely enough' information.

Let V be the space of all possible invariants and $W \subset V$ a subset. A set $X = \{x_1, ..., x_n\} \subset V$ is said to be redundant if there exists an analytic function f such that $x_n = f(x_1, ..., x_{n-1})$. Equivalently, X is non-redundant if

$$f(x_1, ..., x_n) = 0 \quad \Longrightarrow \quad f \equiv 0\,. \tag{1}$$

The set X is complete in W if it generates W with analytic functions, i.e. W is the image of X under analytic functions

$$\{x \in V \mid x = f(x_1, ..., x_n) \text{ for some analytic } f\} = W \,. \tag{2}$$

Consider the set

$$\{t_k \doteq \mathrm{Tr}(D^k), t_{k+1} \doteq \mathrm{Tr}(D^{k+1}), t_{k+2} \doteq \mathrm{Tr}(D^{k+2})\} \tag{3}$$

of Euclidean traces of powers of the diffusion tensor D. For any $k \in \mathbb{N}$ this set is a complete and non-redundant set of invariants by virtue of the Cayley-Hamilton theorem. The example simultaneously illustrates the non-uniqueness property of a tractometric framework. We may arbitrarily choose traces of any three consecutive powers of D to construct a non-redundant complete invariant set. Alternatively we may replace such trace-triples by eigenvalue-triples $\{\lambda_1, \lambda_2, \lambda_3\}$ of D, or by their combinations $\{\mathrm{FA}, \mathrm{MD}, \mathrm{RD}\}$, known as the fractional anisotropy, mean diffusivity and radial diffusivity [3]. These invariants will be refered to as *diffusion invariants*, since they explicitly relate to apparent diffusion properties.

Besides diffusion invariants we wish to include differential properties, taking into account local information in the neighbourhood of a track, i.e. *geometry*. We will make use of the so-called *curvature invariants*, constructed from the Riemann curvature tensor. This tensor determines geodesic deviation [1,31,32,34], and in the context of geodesic tractography it expresses the tendency of nearby tracks to cohere or repel due to local inhomogeneities in the diffusion tensor field. In our Riemannian approach the diffusion tensor D, with components D^{ij} $(i, j = 1, 2, 3)$ relative to a Cartesian coordinate basis, is stipulated to be proportional to the dual Riemannian metric g^{-1}, with components[1] g^{ij}, i.e. $D^{ij} \propto g^{ij}$, so as to 'geometrize away' local diffusivity patterns in the data [11,16,20,25]. Anisotropic water diffusion is then incorporated as intrinsic geometry representing isotropic diffusion in a curved, Riemannian space.

The Riemann curvature tensor R^i_{jkl} is constructed from second order derivatives of the metric tensor, as follows :

$$R^i_{jkl} = \partial_k \Gamma^i_{jl} - \partial_l \Gamma^i_{jk} + \Gamma^i_{km} \Gamma^m_{jl} - \Gamma^i_{lm} \Gamma^m_{jk} \,, \tag{4}$$

where

$$\Gamma^i_{jk} = \frac{1}{2} g^{im} \left(\partial_j g_{km} + \partial_k g_{mj} - \partial_m g_{jk} \right) , \tag{5}$$

and in which ∂_k stands for $\partial / \partial x^k$. Contraction of the Riemann tensor results in the covariant Ricci curvature tensor as well as its mixed and contravariant forms by raising indices with the help of the inverse metric tensor:

$$R_{ij} = R^k_{ikj} \quad R^i_j = g^{ik} R_{kj} \quad \text{and} \quad R^{ij} = g^{ik} g^{jl} R_{kl} \,. \tag{6}$$

[1] Super-/subscripts denote contra-/covariant indices, to which Einstein summation convention applies, i.e. each pair of identical sub- and superscript implies a summation over the corresponding 'dummy' index.

In turn, the Ricci tensor induces a set of curvature invariants given by

$$\{r_1 \doteq R^i_i, r_2 \doteq R^i_j R^j_i, r_3 \doteq R^i_j R^j_k R^k_i\}, \tag{7}$$

in which r_1 is known as the Ricci scalar [8,9]. Thus the set $\{r_1, r_2, r_3\}$ is composed of traces of powers of the matrix with entries R^i_j. By similar arguments as for the diffusion invariants we may instead consider the set $\{\mu_1, \mu_2, \mu_3\}$ of eigenvalues of this matrix.

In either case we prefer the use of eigenvalues $\{\lambda_1, \lambda_2, \lambda_3\}$ and $\{\mu_1, \mu_2, \mu_3\}$ over traces $\{t_1, t_2, t_3\}$, respectively $\{r_1, r_2, r_3\}$, because the former are commensurable and of the same order of magnitude, unlike the latter. Diffusion and curvature invariants provide two complete, non-redundant sets of three invariants each, which we can use separately or jointly as our tractometric invariants of choice.

Given a tractogram of a putative bundle between two anatomical ROIs, we wish to prune it based on the above sets of diffusion and/or curvature invariants by removing 'incoherent' tracks one-by-one in some hierarchical fashion. Consider a bundle of N tracks, each with K associated (averaged) invariants. We thus obtain a correspondence between tracks and points in $\mathbb{R}^K$, yielding a point cloud for the bundle of interest. This point cloud is subsequently pruned with the help of an outlier detection method [5,18,21]. We found that the particular choice of outlier detection method had little effect on the outcome, and chose to use the *isolation forest* algorithm [21]. This algorithm assigns to each cloud point an 'incoherence' score between 0 and 1, indicating its likeliness to be an anomaly within the considered set. A score of 0.5 is then a natural threshold for a particular sample to be labelled as either outlier (score > 0.5) or not (score ≤ 0.5). This threshold could subserve a baseline configuration from which to initialise further pruning by more informed means, possibly involving human interaction. The incoherence score imposes a ranking onto tracks, allowing them to be hierarchically removed one-by-one, cf. Fig. 1.

3 Experiments

In the following experiments we illustrate the tractometric filtering framework on a DWI dataset from the Human Connectome Data Project (dataset "WU-Minn HCP Data-1200 Subjects": subject 100307; TE/TR/echo spacing 89.5/5520.0/0.78 ms; b = 2000 s/mm^2), as well as on a clinical dataset acquired with a Philips Achieva 3T MRI scanner (b = 1500 s/mm^2, 50 diffusion-weighting directions, six b = 0 s/mm^2 images, 2 mm isotropic voxel size, TE/TR/echo spacing 87/8000/0.2 ms). DTI tensors are computed using weighted least squares via the Dipy library in Python [12]. Throughout our experiments the defining Gram matrix for the Riemannian metric is the adjugate of the diffusion tensor D, i.e. $g_{ij} = \det(D) D^{\text{inv}}_{ij}$, cf. Fuster et al. [11].

We consider two experiments, the corticospinal tract (CST) on the HCP dataset and the arcuate fasciculus (AF) on the clinical dataset, to illustrate that the proposed method has a qualitatively different behaviour on a per bundle

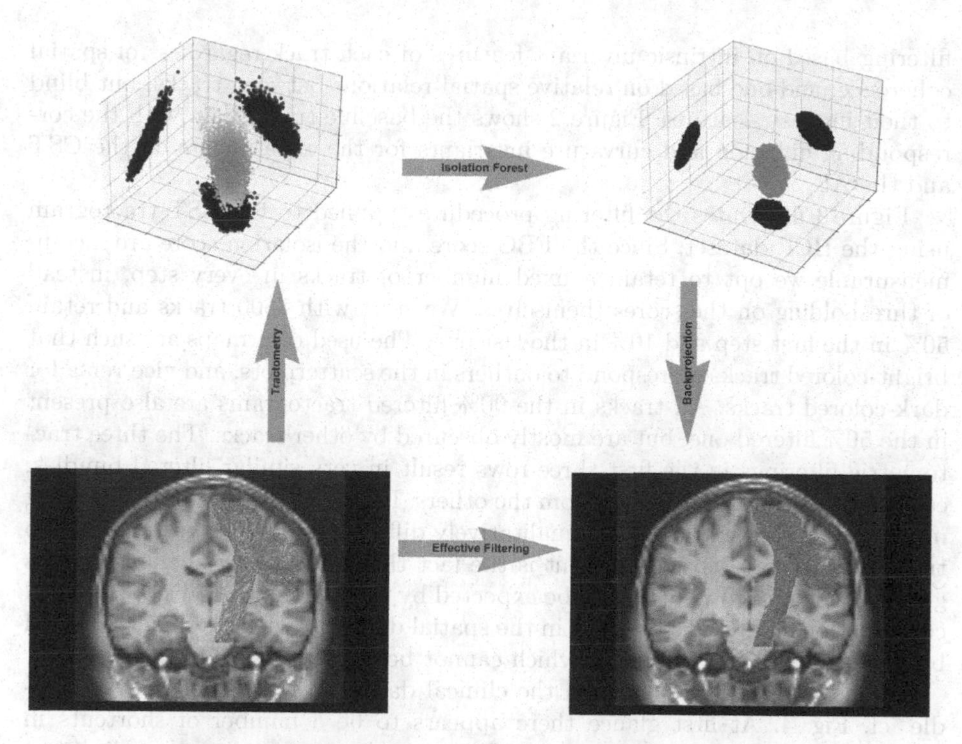

Fig. 1. An example of the filtering method for the corticospinal tract (CST) using the diffusion invariants. **Top left:** Scatterplot of the three invariants $\{\lambda_1, \lambda_2, \lambda_3\}$ for each of the 7500 tracks in the tractogram, color-coded according to the isolation score as determined by the isolation forest algorithm. Each invariant $\{\lambda_1, \lambda_2, \lambda_3\}$ is projected onto the coordinate planes in black. **Top right:** Scatterplot of the invariants after removing all points with isolation score larger than 0.5. **Bottom:** Tractograms corresponding to the scatterplots above, with the unfiltered tractogram on the left and the filtered one on the right. The arrow 'Effective Filtering' is defined by virtue of the other three, 'Tractometry', 'Isolation Forest' and 'Backprojection'.

basis. Furthermore, these two bundles are of clinical interest to a collaborating neurosurgeon, clarifying the use of a clinical dataset in the AF-experiment. Both experiments are set up according to the following structure. We depart from a probabilistic streamline tractogram computed using the iFOD2 algorithm included in *tckgen* in MRtrix3 [36] following the data processing pipeline described in [23]. We then apply four different filterings, three of which are based on our tractometric invariants, viz. the set of three diffusion invariants, the set of three curvature invariants, respectively and their union (containing all six invariants). All track-wise invariants are computed by sampling the voxel-wise invariants along the curve using trilinear interpolation. The fourth filtering, introduced as comparison reference, relies on a scalar measure based on the relative spatial arrangement of tracks in a bundle, the Fiber-to-Bundle Coherence (FBC) [22]. This provides us with a comparison between our tractometric filtering based on intrinsic invariant features of each track regardless of spatial coherence, and one based on relative spatial relations between tracks, but blind to their intrinsic features. Figure 2 shows the baseline tractogram with the corresponding diffusion and curvature invariants for the experiments on the CST and the AF.

Figure 3 illustrates the filtering procedures applied to the CST tractogram using the HCP dataset. Since the FBC score and the isolation score are incommensurable we opt to retain a fixed number of tracks in every step, instead of thresholding on the scores themselves. We start with 7500 tracks and retain 50% in the first step and 10% in the last one. The used colormaps are such that bright-colored tracks correspond to outliers in the scatterplots, and vice versa for dark-colored tracks. All tracks in the 90% filtered tractograms are also present in the 50% filtered one, but are mostly obscured by other tracks. The three tractometric filterings in the first three rows result in very similar filtered bundles, each one differing only slightly from the others. Interestingly, the spatial pruning method (fourth row) displays a qualitatively different behaviour with respect to track elimination. Most prominent is the fact that the fanning of the bundle is gradually destroyed (which is to be expected by the fourth method's operational construction enforcing coherence in the spatial domain), whereas this is retained by using tractometric filtering (which cannot be expected a priori).

In the second experiment, on the clinical dataset, we consider the AF bundle, cf. Fig. 4. At first glance there appears to be a number of shortcuts in the tractograms. This behaviour is reflected in the scatterplots, revealing two point clusters. In all cases, tractometric filtering removes first the smaller cluster, thereby eliminating the shortcuts from the tractogram, in correspondence with spatial FBC-based pruning. In this case, spatial coherence of tracks apparently correlates strongly with intrinsic tractometric features.

4 Discussion

We have proposed a method for filtering diffusion MRI tractograms. Based on the assumption that anatomical (sub)bundles are internally structurally coherent, we have constructed characteristic trackwise scalar invariants, combined into complete and non-redundant sets so as to capture all degrees of freedom of a given differential order. For each track, the associated set of invariants defines coordinates of a point in a feature space. The tractogram of the bundle of interest is then effectively filtered by removing outliers from the corresponding point cloud in this space.

Experiments indicate that this method may be used in conjunction to existing methods, such as filters based on spatial coherence of the tracks. It depends on the anatomical bundle of interest which method is most appropriate. Whereas spatial filters will tend to eliminate spatial outliers by construction, tractometric filters are blind to spatial configurations and driven by intrinsic characteristics of individual tracks. For instance, sub-bundles or branches of a fanning bundle may or may not have distinct intrinsic characteristics. Future research should

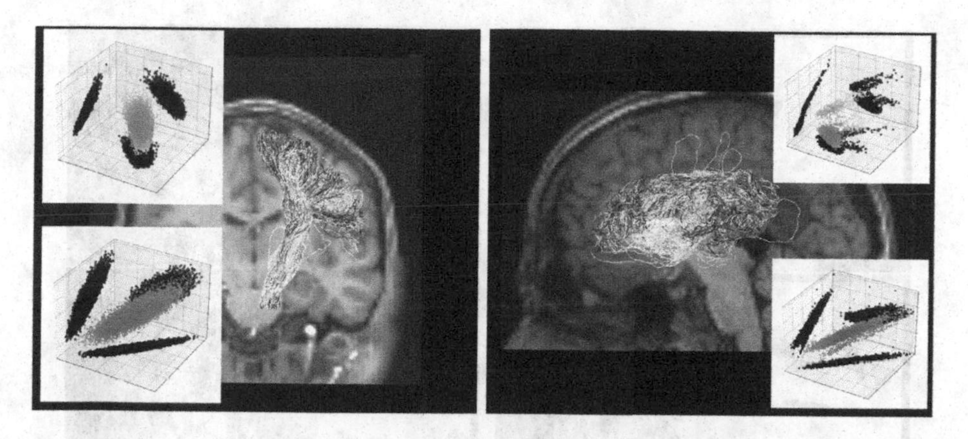

Fig. 2. The baseline tractograms for experiments on the CST (left) and AF (right). The former experiment is performed on the HCP dataset and the latter on the clinical dataset. The scatterplots depict the diffusion invariants (red-yellow) and the curvature invariants (light-dark blue). The tractograms are color-coded according to the isolation score ranking corresponding to the diffusion invariants. While we could have chosen to use the curvature invariants or the union of both sets, this is merely a practical choice to illustrate the behaviour of scalar features in the tractogram. (Color figure online)

help to identify the 'elementary' anatomical (sub-)bundles for which a particular method is appropriate. A clustering method applied prior to pruning in the case of multi-modal point clouds might be instrumental. Experiments with multiple (sub-)bundles in the same subject as well as studying the inter-subject variability of this method are interesting avenues for further research.

Interestingly, by employing distinct sets of independent invariants, viz. diffusion and curvature invariants, one may a priori expect a difference in the outcome. However, the experiments show that, at least qualitatively, using each in isolation or combining both, hardly affects the result. This is an indication that the premise of microstructural coherence of an anatomically plausible bundle might be correct, since that would reflect itself in macroscopic coherence of invariants regardless of their precise nature.

Our framework can be extended in multiple ways. One may add new sets of invariants and consider various combinations. Considering higher order derivatives of the metric tensor field can systematically add complexity to the sets of invariants. Moreover, extrinsic spatial (such as FBC) and intrinsic tractometric features may be combined in a hybrid method. Other integral or projective operators to construct invariant functionals, besides the simple case of the mean operator used in this paper, may be considered. In particular one may define point-wise invariant functionals for each tract, leading to more complex, high-dimensional descriptors of a tractogram. Our framework may find applicability beyond DTI, e.g. using RISH features for the spherical harmonic representation of fiber orientation density functions [24].

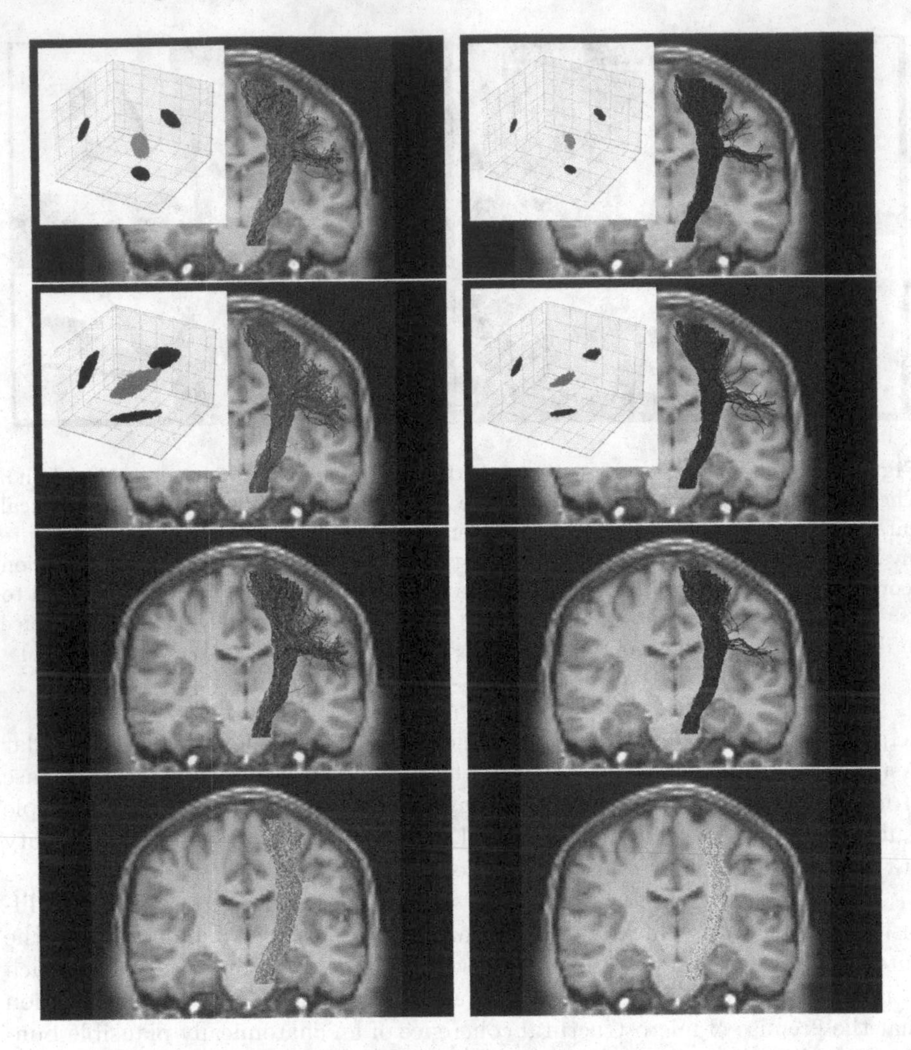

Fig. 3. The effect of filtering the CST tractograms, see Fig. 2, with different sets of invariants as well as using a spatial filter. Each row represents a different filter and each column a threshold by which to prune the tractogram (expressed in percentage of the total number of tracks). **Columns:** Filtered tractograms with 50%, resp. 10% of the tracks remaining. **Rows 1–4:** Diffusion invariants and corresponding scatterplot; Curvature invariants and corresponding scatterplot; Combined diffusion and curvature invariants; Fiber-to-Bundle Coherence.

Secondly, even though tractometry is only applied to streamline tractograms in our experiments, the curvature invariants would have a more transparent interpretation in a Riemannian geodesic framework, for which they were designed in the first place in the context of uncertainty quantification given data perturbations (noise and end-point conditions). Interpretability of features in terms of

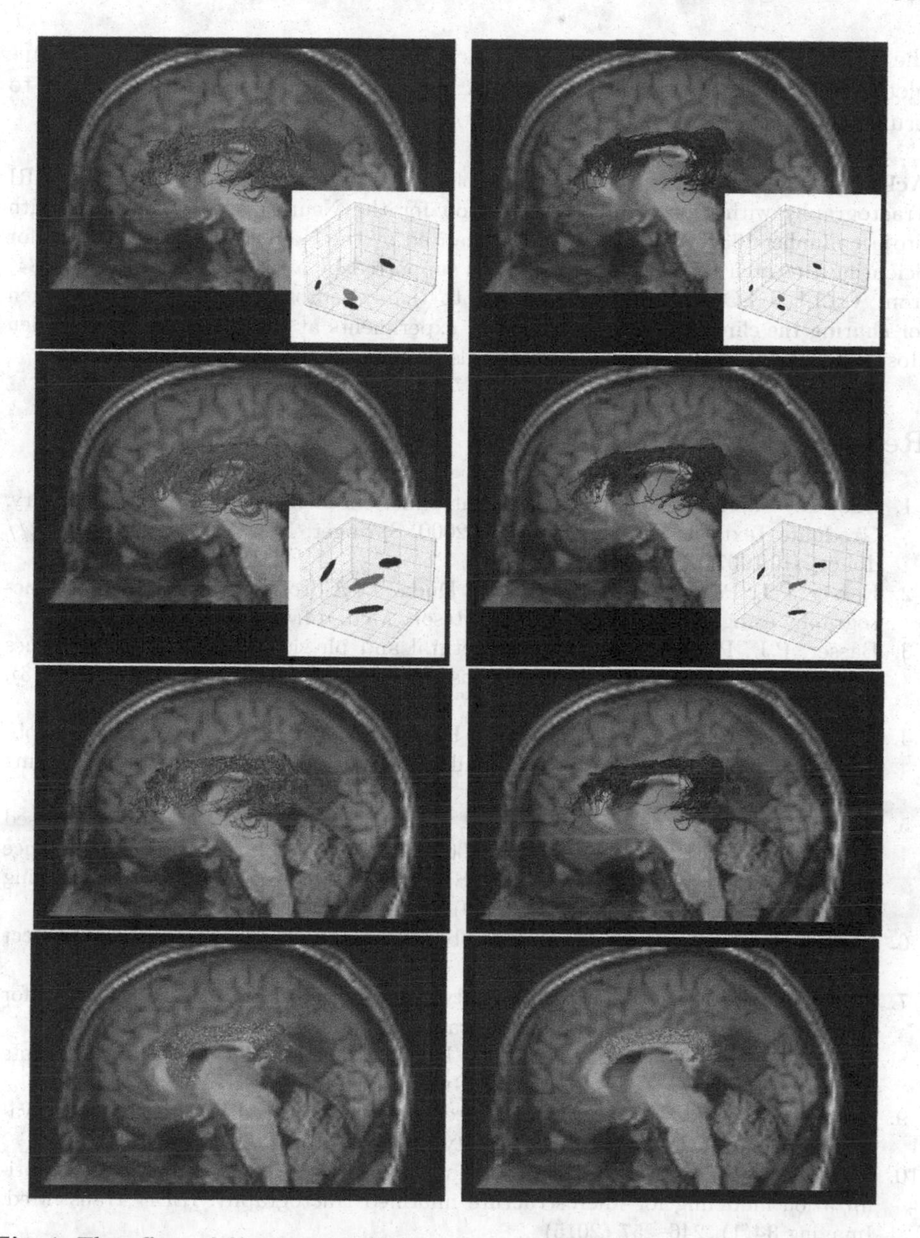

Fig. 4. The effect of filtering the AF tractograms with different sets of invariants as well as using a spatial filter. Each row represents a different filter and each column a threshold by which to prune the tractogram (expressed in percentage of the total number of tracks). Filtered tractograms with 50%, resp. 10% of the tracks remaining. **Rows 1–4:** Diffusion invariants and corresponding scatterplot; Curvature invariants and corresponding scatterplot; Combined diffusion and curvature invariants; Fiber-to-Bundle Coherence.

the specific details of a tractography method is, however, subordinate to completeness. Complete sets of tractometric invariants can therefore be applied to prune any tractogram, regardless of its operational definition.

Acknowledgements. This work is part of the research programme "Diffusion MRI Tractography with Uncertainty Propagation for the Neurosurgical Workflow" with project number 16338, which is (partly) financed by the Netherlands Organisation for Scientific Research (NWO). This work was supported by a research grant (00028384) from VILLUM FONDEN. We would like to thank neurosurgeon Geert-Jan Rutten for sharing the clinical dataset used in our experiments at the Elisabeth TweeSteden Hospital (ETZ) in Tilburg, The Netherlands, and for fruitful discussions.

References

1. Bao, D., Chern, S.S., Shen, Z.: An Introduction to Riemann-Finsler Geometry. Graduate Texts in Mathematics, vol. 2000. Springer, New York (2000). https://doi.org/10.1007/978-1-4612-1268-3
2. Basser, P.J., Pajevic, S., Pierpaoli, C., Duda, J., Aldroubi, A.: In vivo fiber tractography using DT-MRI data. Magn. Reson. Med. **44**(4), 625–632 (2000)
3. Basser, P.J., Pierpaoli, C.: Microstructural and physiological features of tissues elucidated by quantitative-diffusion-tensor MRI. J. Magn. Reson. Series B **111**(3), 209–219 (1996)
4. Behrens, T.E.J., Berg, H.J., Jbabdi, S., Rushworth, M.F.S., Woolrich, M.W.: Probabilistic diffusion tractography with multiple fibre orientations: what can we gain? Neuroimage **34**(1), 144–155 (2007)
5. Breunig, M.M., Kriegel, H.P., Ng, R.T., Sander, J.: LOF: identifying density-based local outliers. In: Proceedings of the 2000 ACM SIGMOD International Conference on Management of Data, pp. 93–104. SIGMOD 2000, Association for Computing Machinery, New York, NY, USA (2000)
6. Chamberland, M., et al.: Tractometry-based anomaly detection for single-subject white matter analysis (2020)
7. Chamberland, M., et al.: Dimensionality reduction of diffusion MRI measures for improved tractometry of the human brain. Neuroimage **200**, 89–100 (2019)
8. Coley, A.A., MacDougall, A., McNutt, D.D.: Basis for scalar curvature invariants in three dimensions. Class. Quantum Gravity **31**(23), 235010 (2014)
9. Coley, A., Hervik, S., Pelavas, N.: Spacetimes characterized by their scalar curvature invariants. Class. Quantum Gravity **26**(2), 025013 (2009)
10. Daducci, A., Dal Palù, A., Lemkaddem, A., Thiran, J.P.: Commit: convex optimization modeling for microstructure informed tractography. IEEE Trans. Med. Imaging **34**(1), 246–257 (2015)
11. Fuster, A., Dela Haije, T., Tristán-Vega, A., Plantinga, B., Westin, C.F., Florack, L.: Adjugate diffusion tensors for geodesic tractography in white matter. J. Math. Imaging Vis. **54**(1), 1–14 (2016)
12. Garyfallidis, E., et al.: Dipy, a library for the analysis of diffusion MRI data. Front. Neuroinf. 8 (2014)
13. Garyfallidis, E., Brett, M., Correia, M.M., Williams, G.B., Nimmo-Smith, I.: QuickBundles, a method for tractography simplification. Front. Neurosci. **6**, 175 (2012)

14. Garyfallidis, E., et al.: Recognition of white matter bundles using local and global streamline-based registration and clustering. Neuroimage **170**, 283–295 (2018)
15. Hao, X., Whitaker, R.T., Fletcher, P.T.: Adaptive Riemannian metrics for improved geodesic tracking of white matter. In: Székely, G., Hahn, H.K. (eds.) IPMI 2011. LNCS, vol. 6801, pp. 13–24. Springer, Heidelberg (2011). https://doi.org/10.1007/978-3-642-22092-0_2
16. Hao, X., Zygmunt, K., Whitaker, R.T., Fletcher, P.T.: Improved segmentation of white matter tracts with adaptive Riemannian metrics. Med. Image Anal. **18**, 161–175 (2014)
17. Jost, J.: Riemannian Geometry and Geometric Analysis. Universitext, 6th edn. Springer, Berlin (2011). https://doi.org/10.1007/978-3-642-21298-7
18. Knorr, E.M., Ng, R.T., Tucakov, V.: Distance-based outliers: algorithms and applications. VLDB J. **8**(3), 237–253 (2000)
19. Legarreta, J.H., et al.: Filtering in tractography using autoencoders (FINTA). Med. Image Anal. **72**, 102126 (2021)
20. Lenglet, C., Deriche, R., Faugeras, O.: Inferring white matter geometry from diffusion tensor MRI: application to connectivity mapping. In: Pajdla, T., Matas, J. (eds.) ECCV 2004. LNCS, vol. 3024, pp. 127–140. Springer, Heidelberg (2004). https://doi.org/10.1007/978-3-540-24673-2_11
21. Liu, F.T., Ting, K.M., Zhou, Z.H.: Isolation forest. In: 2008 Eighth IEEE International Conference on Data Mining, pp. 413–422 (2008)
22. Meesters, S., et al.: Stability metrics for optic radiation tractography: towards damage prediction after resective surgery. J. Neurosci. Methods **288**, 34–44 (2017)
23. Meesters, S., Rutten, G.J., Fuster, A., Florack, L.: Automated tractography of four white matter fascicles in support of brain tumor surgery. In: 2019 OHBM Annual Meeting, June 6–13 2019, Rome, Italy. Organization for Human Brain Mapping (2019), abstract nr. Th768
24. Mirzaalian, H., et al.: Harmonizing diffusion MRI data across multiple sites and scanners. In: Medical image computing and computer-assisted intervention: MICCAI ... International Conference on Medical Image Computing and Computer-Assisted Intervention 9349, 12–19 October (2015)
25. O'Donnell, L., Haker, S., Westin, C.-F.: New approaches to estimation of white matter connectivity in diffusion tensor MRI: Elliptic PDEs and geodesics in a tensor-warped space. In: Dohi, T., Kikinis, R. (eds.) MICCAI 2002. LNCS, vol. 2488, pp. 459–466. Springer, Heidelberg (2002). https://doi.org/10.1007/3-540-45786-0_57
26. O'Donnell, L., Westin, C.F.: Automatic tractography segmentation using a high-dimensional white matter atlas. IEEE Trans. Med. Imaging **26**, 1562–75 (2007)
27. Rund, H.: The Differential Geometry of Finsler Spaces. Springer, Berlin (1959). https://doi.org/10.1007/978-3-642-51610-8
28. Schober, M., Kasenburg, N., Feragen, A., Hennig, P., Hauberg, S.: Probabilistic shortest path tractography in DTI Using Gaussian process ODE solvers. In: Golland, P., Hata, N., Barillot, C., Hornegger, J., Howe, R. (eds.) MICCAI 2014. LNCS, vol. 8675, pp. 265–272. Springer, Cham (2014). https://doi.org/10.1007/978-3-319-10443-0_34
29. Sebastiani, G., De Pasquale, F., Barone, P.: Quantifying human brain connectivity from diffusion tensor MRI. J. Math. Imaging is. **25**(2), 227–244 (2006)
30. Sengers, R., Florack, L., Fuster, A.: Geodesic uncertainty in diffusion MRI. Front. Comput. Sci. **3**, 718131 (2021)

31. Sengers, R., Florack, L., Fuster, A.: Geodesic tubes for uncertainty quantification in diffusion MRI. In: Feragen, A., Sommer, S., Schnabel, J., Nielsen, M. (eds.) IPMI 2021. LNCS, vol. 12729, pp. 279–290. Springer, Cham (2021). https://doi.org/10.1007/978-3-030-78191-0_22
32. Shen, Y.B., Shen, Z.: Introduction to Modern Finsler Geometry. World Scientific, Singapore (2016)
33. Smith, R.E., Tournier, J.D., Calamante, F., Connelly, A.: Sift2: enabling dense quantitative assessment of brain white matter connectivity using streamlines tractography. Neuroimage **119**, 338–351 (2015)
34. Spivak, M.: Differential Geometry, vol. 1–5. Publish or Perish, Berkeley (1975)
35. St-Jean, S., Chamberland, M., Viergever, M.A., Leemans, A.: Reducing variability in along-tract analysis with diffusion profile realignment. Neuroimage **199**, 663–679 (2019)
36. Tournier, J.D., Smith, R., Raffelt, D., Tabbara, R., Dhollander, T., Pietsch, M., Christiaens, D., Jeurissen, B., Yeh, C.H., Connelly, A.: MRtrix3: a fast, flexible and open software framework for medical image processing and visualisation. Neuroimage **202**, 116137 (2019)
37. Westin, C.-F., Maier, S.E., Khidhir, B., Everett, P., Jolesz, F.A., Kikinis, R.: Image processing for diffusion tensor magnetic resonance imaging. In: Taylor, C., Colchester, A. (eds.) MICCAI 1999. LNCS, vol. 1679, pp. 441–452. Springer, Heidelberg (1999). https://doi.org/10.1007/10704282_48
38. Yeatman, J.D., Dougherty, R.F., Myall, N.J., Wandell, B.A., Feldman, H.M.: Tract profiles of white matter properties: automating fiber-tract quantification. PLoS ONE **7**(11), e49790 (2012)

Author Index

Alemán-Gómez, Yasser 3
Alexander, Daniel C. 77

Baete, Steven H. 89
Barnett, Michael 65
Basler, Lee 89
Blumberg, Stefano B. 77
Boada, Fernando E. 89

Cabezas, Mariano 65
Cai, Weidong 65
Calamante, Fernando 65
Canales-Rodríguez, Erick Jorge 3
Chen, Geng 115
Cuadra, Meritxell Bach 3
Cui, Hui 115

D'Souza, Arkiev 65
Dela Haije, Tom 137
Descoteaux, Maxime 125

Epstein, Sean C. 77

Filipiak, Patryk 89
Florack, Luc 137
Fuster, Andrea 137

Girard, Gabriel 3
Goodwin-Allcock, Tobias 101
Graham, Mark 38
Gray, Robert 101
Guerreri, Michele 38

Jahanshad, Neda 50
Jakab, András 3
Jodoin, Pierre-Marc 125
Jones, Derek K. 26

Kebiri, Hamza 3

Legarreta, Jon Haitz 125
Legouhy, Antoine 38
Lim, Jason P. 77
Liu, Dongnan 65

Ma, Jiquan 115
Mancini, Matteo 26
McEwen, Jason 101

Nachev, Parashkev 101
Narayan, Neil 77
Nir, Talia M. 50

Palombo, Marco 26, 77
Peigneux, Philippe 38
Petit, Laurent 125
Placantonakis, Dimitris G. 89

Schneider, Walter 89
Sengers, Rick 137
Shepherd, Timothy 89
Slator, Paddy J. 77
Spears, Tyler A. 14
Stee, Whitney 38

Tang, Zihao 65
Thomas Fletcher, P. 14
Thompson, Paul M. 50

Villalón-Reina, Julio E. 50
Villemonteix, Thomas 38

Wang, Chenyu 65
Wang, Xinyi 65

Xu, Pengbo 115

Yin, Haoran 115
Yu, Thomas 3

Zhang, Hui 38, 101
Zuccolotto, Anthony 89

Printed in the United States
by Baker & Taylor Publisher Services

Printed in the United States
by Baker & Taylor Publisher Services